Biochemistry Research Trends

Peroxidases

Biochemical Characteristics, Functions and Potential Applications

BIOCHEMISTRY RESEARCH TRENDS

Additional books in this series can be found on Nova's website
under the Series tab.

Additional e-books in this series can be found on Nova's website
under the e-book tab.

PROTEIN SCIENCE AND ENGINEERING

Additional books in this series can be found on Nova's website
under the Series tab.

Additional e-books in this series can be found on Nova's website
under the e-book tab.

BIOCHEMISTRY RESEARCH TRENDS

PEROXIDASES

BIOCHEMICAL CHARACTERISTICS, FUNCTIONS AND POTENTIAL APPLICATIONS

LUCAS BOGAERT
AND
NOAH COPPENS
EDITORS

New York

For permission to use material from this book please contact us:
Telephone 631-231-7269; Fax 631-231-8175
Web Site: http://www.novapublishers.com

NOTICE TO THE READER

Additional color graphics may be available in the e-book version of this book.

Library of Congress Cataloging-in-Publication Data

Library of Congress Control Number: 2013942222

ISBN: 978-1-62808-261-6

Published by Nova Science Publishers, Inc. † New York

Contents

Preface

In this book, the authors present topical research in the study of the biochemical characteristics, functions and potential applications of peroxidases. Topics discussed include the unique properties and biotechnological potential of ligninolytic peroxidases; the functions and regulation during pathogenesis of plant peroxidases; phospholipids and peroxidases in mitochondrial apoptosis; glutathione peroxidases in stem, cancer, and cancer stem cells; biochemical responses of aquatic macrophytes to two abiotic stresses; role of ascorbate peroxidase in antioxidant protection; and the route, dose and duration of exposure to cadmium-relevance to oxidative stress induction.

Chapter I – Ligninolytic fungi (white rot and litter-decomposing fungi) are a physiological rather than taxonomic group connecting fungal strains which are capable of extensively degrading lignin. The unique physiological and biochemical properties permit these fungi to transform and mineralize of wide range of organopollutants with structural similarities to lignin. They produce a complex of extracellular oxidative enzymes, including lignin peroxidase, manganese peroxidase, versatile peroxidase, and laccase, which are involved in degradative processes. This enzyme complex is produced by all ligninolytic fungi but its pattern is dependent on fungal species and cultivation conditions. Lignin peroxidase catalyzes oxidative cleavage of C-C bonds and ether (C-O-C) bonds in non-phenolic aromatic substrates with high redox potential. Manganese peroxidase oxidizes Mn^{2+} to Mn^{3+}, which facilitates the degradation of phenolic compounds or, in turn, oxidizes a second mediator for the breakdown of non-phenolic compounds. Versatile peroxidase is hybrid of lignin peroxidase and manganese peroxidase combining the catalytic possibilities of those two peroxidases. This review presents the most important molecular and catalytic properties of ligninolytic peroxidases including molecular structure, substrate specificity, optimal pH values and temperature, influence of effectors, catalytic activity towards lignin and xenobiotics. The features of catalytic mechanisms of these enzymes are comparing. The possible functions of ligninolytic peroxidases and supposed dependence the fungal degrading properties on pattern and properties to produce various peroxidases are discussed. Finally the potential of biotechnological use of these enzymes is under consideration.

Chapter II – Peroxidase (donor: H_2O_2oxidoreductase, EC 1.11.1.7) (PO) belongs to a class of widespread and vital enzymes. One of its specific features is the multiplicity of its molecular isoforms exhibiting various functions], and the highly variable genes encoding them are controlled by various *cis*-elements and *trans*-factors that are responsible for their differentiated expression activity depending on environmental conditions and stages of ontogenesis. POs are localizing in cytoplasm, plasma membrane or secrete outside of cell. In cell wall POs presence in the ionic-bound or covalent bound fraction of proteins, and even freely circulated in apoplast. POs are key enzymes of lignification capable of binding with polysaccharides of plant and (or) pathogens. An important physiological sense of these reactions is the strengthening of the cell wall and construction on the pathogens' paths of physical barriers consisting of lignin- and suberin-containing polymers, which also possess high toxicity. This defense reaction is developing more or less automatically, thus these barriers are formed just in the zone of pathogen penetration and concurrently with the beginning of the active pathogen expansion into the tissues of the host plant. However, mechanisms of development of these events are still unclear. Expression of some genes encoding different isoPOs are regulated by pathogens (and their metabolites), elicitors, and stress-related hormones and hormone-like compounds, such as salicylic and jasmonic acids, ethylene and others.

Chapter III – Phospholipids constitute an essential part of cellular membranes, in particular, inner and outer membranes of mitochondria. They influence the function of mitochondrial proteins and cytosolic proteins associates with mitochondria. Phospholipids regulate the assembly and maintenance of the proteins embedded in the membranes, they modulate enzymes' ox-red properties, electron-transfer properties, kinase activity, and peroxidase activity. Phospholipids thus play a crucial role in all cellular processes such as protein biogenesis, protein import into mitochondria, energy production, and apoptosis. Over the last decade, a significant amount of research efforts has been devoted to mitochondrial apoptosis, in particular, early stages of apoptosis that precede mitochondrial permeability transition pore (PTP) opening and the release of proapoptotic factors into the cytoplasm, specifically cytochrome c (cyt c). Cyt c is a soluble protein located in the intermembrane space of mitochondria whose primary function is the electron shuttle between complexes III and IV of the respiratory chain. Cyt c is attached to the inner mitochondrial membrane (IMM) through a phospholipid molecule, cardiolipin (CL). It was demonstrated that cyt c, whose peroxidase function is negligible in a free state, exhibits a significant peroxidase activity upon binding to CL. This cyt c-CL complex catalyzes the peroxidation of CL. Peroxidized CL stimulates the formation of the permeability pore in the membranes and the release of cyt c into the cytoplasm through these pores. The increase of the peroxidase activity of cyt c in the CL-bound state is thus considered to be a prerequisite for apoptosis. Cells have also developed anti-apoptotic mechanisms. One of these anti-apoptotic functions is executed by a mitochondrial phospholipid hydroperoxide glutathione peroxidase or GPx4, a protein that defends cells from oxidative injury, attenuates cardiac disfunction, and participates in sperm development. GPx4 plays an essential role in preventing cyt c release from mitochondria. Specifically, it reduces oxidized phospholipids precluding cyt c detachment from CL and consequently suppressing apoptosis. The authors will describe the structure of cyt c and GPx4 as well as the catalytic cycles of the two peroxidases. The authors will highlight the recent advances and unresolved issues regarding biochemical mechanisms underlying cyt c-CL interactions. The authors will discuss the structural characteristics of cyt c-CL interactions responsible for the tight binding, unfolding of cyt c, and the enhancement of its peroxidase activity. The authors will also discuss how peroxidized CL triggers PTP opening and how the balance between the two peroxidases, CL-bound cyt c and GPx4, affects the early stages of mitochondrial apoptosis.

Chapter IV – Oxidative stress, as a disturbance in the equilibrium between free radicals (FR), reactive oxygen species (ROS), and endogenous antioxidant defense mechanisms is caused by an imbalance between the production of reactive oxygen and a biological system's ability to readily detoxify the reactive intermediates or easily repair the resulting damage. Mounting evidence has implicated that oxidative stress is involved in various physiological and pathological processes, including DNA damage, proliferation, cell adhesion, and survival of cancer cells. Glutathione peroxidases (GPxs) (EC 1.11.1.9), which can reduce lipid hydroperoxides to their corresponding alcohols and free hydrogen to water, are enzyme family with peroxidase activity whose main biological roles are to protect the organism from oxidative damage. So far 8 sub-members of GPxs have been identified in humans, i.e. GPx 1–8. They all can react with H_2O_2 and soluble fatty acid hydroperoxides. A large number of publications have demonstrated that GPxs have significant roles in different stages of carcinogenesis. In this book chapter, the authors will review recent study progress of GPxs in stem, cancer and cancer stem cells. This knowledge will shed light on mechanisms that lead to better modes of cancer treatment.

Chapter V – Exposure to stress can lead to the disruption of cellular, molecular and physiological processes in plants. Plants generate ROS to cope with different forms of abiotic stresses. However, if a plant is exposed to stress conditions for a considerable period, the excess generation of ROS affects its radical scavenging ability in the long run. Thus the effects of environmental variability of a plant are reflected in its growth, reproduction and colonization. Aquatic plants are considered unchangeable biological filters and play an important role in the proper functioning of aquatic ecosystem. They often encounter abiotic stress of varying magnitudes and acquire with different adaptive characters. In this chapter, the authors will be focusing on the oxidative responses of aquatic macrophytes to various abiotic stresses with a special concern on flow velocity and redox states.

Chapter VI – The change in activities and numerous molecular forms of antioxidant enzymes is one of the parameters of plant biological tolerance against constantly changing environmental conditions. A group of genes involved in the synthesis of ascorbate is used in genetic engineering for the improvement of plant tolerance to biotic and abiotic stresses. The level of H_2O_2 in plant cells is controlled by two enzymes: catalase and ascorbate peroxidase (APX), though the key role in decomposition of H_2O_2 belongs to APX. Ascorbate peroxidase converts H_2O_2 into H_2O and O_2. The APX activity and electrophoresis spectra of its isoenzyme content have been investigated in plants subjected to abiotic (soil drought) and biotic (virus infections) stresses. The analysis of electrophoretic spectra of four isoforms in wheat leaves at the beginning of drought (flowering phase) revealed four isoforms. The amounts of APX isoforms increased to 7 during the wax ripeness phase both in irrigated plants and plants subjected to drought. The long-term drought led to the increase in the intensity of high-molecular isoforms and emergence of three additional isoforms of the enzyme. In this instance, the APX activity level remained higher in drought tolerant wheat genotypes compared with the sensitive ones. At the same time, in both tolerant and sensitive varieties, enhanced APX activity compared with the control variants was observed at the end of drought, which is apparently related to the synthesis of new molecules of the enzyme (i.e. it is stipulated by the changes of genome expression under stress). The immune enzymatic method revealed the presence of CMV virus in *C. sativus* L. plants and TMV in *L. esculentum* Mill. plants. The obtained results show significant differences in APX activities in all virus infected leaves compared with controls. The effect of the viral infection caused approximately a 1.6-fold increase of the APX activity in *L. esculentum* Mill. and a 2.1-fold increase in *C. sativus* L. compared with corresponding controls. Pathogenesis was accompanied by an increase in the quantity of APX isoforms, which reached 5 in leaves of *L. esculentum* Mill. and by the formation of 2 new isoforms and intensification of the enzyme electrophoresis spectra in infected leaves of *C. sativus* L. compared with the control variants. Thus, an increase in ascorbate peroxidase activity and *de novo* synthesis of its numerous molecular forms under stress conditions testify its involvement in the antioxidant defense system.

Chapter VII – Cadmium (Cd) is a toxic metal currently ranked as seventh (out of 275) on ATSDR priority list of hazardous substances and it is considered as one of the most important occupational and environmental pollutants. This carcinogenic metal causes adverse effects in various tissues, particularly kidneys after prolonged exposure. However, the mechanisms of these toxic effects are still not completely understood. Literature data indicate different mechanisms of cadmium toxicity, oxidative stress being one of the most important. Experiments confirm that Cd, metal with no redox potential, can indirectly induce the production of reactive oxygen species (ROS) by affecting antioxidant enzymes (superoxide dismutase, catalase, glutathione peroxidase, glutathione reductase, glutathione S-transferase), nonenzymatic antioxidants (glutathione, total sulfhydryl groups, antioxidative vitamines), and Fenton metals, iron and copper. This chapter summarizes our investigations on cadmium-induced oxidative stress conducted on three different animal species: rabbits, mice and rats. Application of a single oral dose of 50 mg Cd/kg b.w. in rabbits showed increased levels of lipid peroxidation in blood 4-6 hours after Cd intoxication indicating early development of oxidative stress in conditions of acute Cd poisoning. Experiments carried out on mice showed that single oral exposure to 20 mg Cd/kg b.w. resulted in intensive ROS production in liver while subacute oral dose of 10 mg Cd/kg b.w. induced more pronounced effect in kidneys, thus indicating different effect of Cd intoxication depending on duration of Cd exposure. Further experiments performed on rats provided strong evidence that Cd can induce oxidative stress in both blood and liver after oral (30 mg Cd/kg b.w.) and intraperitoneal (1.5 mg Cd/kg b.w.) acute intoxication, although intraperitoneal administration showed more pronounced negative effects. Results of these studies demonstrate that oxidative stress is an important mechanism of Cd toxicity and that Cd ability to induce oxidative stress in different tissues depends on the route, dose and duration of exposure to this toxic metal.

In: Peroxidases
Editors: Lucas Bogaert and Noah Coppens

ISBN: 978-1-62808-261-6

Chapter I

Ligninolytic Peroxidases: Unique Properties and Biotechnological Potential

N. Pozdnyakova[1*], K. Wlizlo[2], K. Szalapata[2], A. Jarosz-Wilkolazka[2] and O. Turkovskaya[1]

[1]Institute of Biochemistry and Physiology of Plants and Microorganisms, Russian Academy of Sciences, Saratov, Russia

[2]Department of Biochemistry, Maria Curie-Sklodowska University, Lublin, Poland

Abstract

Ligninolytic fungi (white rot and litter-decomposing fungi) are a physiological rather than taxonomic group connecting fungal strains which are capable of extensively degrading lignin. The unique physiological and biochemical properties permit these fungi to transform and mineralize of wide range of organopollutants with structural similarities to lignin. They produce a complex of extracellular oxidative enzymes, including lignin peroxidase, manganese peroxidase, versatile peroxidase, and laccase, which are involved in degradative processes. This enzyme complex is produced by all ligninolytic fungi but its pattern is dependent on fungal species and cultivation conditions. Lignin peroxidase catalyzes oxidative cleavage of C-C bonds and ether (C-O-C) bonds in non-phenolic aromatic substrates with high redox potential. Manganese peroxidase oxidizes Mn^{2+} to Mn^{3+}, which facilitates the degradation of phenolic compounds or, in turn, oxidizes a second mediator for the breakdown of non-phenolic compounds. Versatile peroxidase is hybrid of lignin peroxidase and manganese peroxidase combining the catalytic possibilities of those two peroxidases. This review presents the most important molecular and catalytic properties of ligninolytic peroxidases including molecular structure, substrate specificity, optimal pH values and temperature, influence of effectors, catalytic activity towards lignin and xenobiotics. The features of catalytic mechanisms of these

* Corresponding author: nataliapozdnyakova@yahoo.com.

enzymes are comparing. The possible functions of ligninolytic peroxidases and supposed dependence the fungal degrading properties on pattern and properties to produce various peroxidases are discussed. Finally the potential of biotechnological use of these enzymes is under consideration.

Introduction

Degradation of lignin is a key process for carbon recycling in land ecosystems. Degradation of this recalcitrant aromatic polymer is caused in nature by special group of high fungi, so called ligninolytic fungi. Ligninolytic fungi are rather physiological than systematic group of wood-inhabiting (white rot) and related them terricolous (litter-decomposing) fungi which have ability to depolymerize and mineralize lignin through a process that was defined as an enzymatic combustion [1].

These fungi have received worldwide attention because of their potential biotechnological use in degradation of lignin and humic substances [2], in decolourization of dye-containing effluents [3, 4], and in detoxifying recalcitrant environmental pollutants, such as dioxins, chlorinated compounds [5, 6], pharmaceutical compounds [7], endocrine-disrupting compounds [8-10], trinitrotoluene [11, 12], polycyclic aromatic hydrocarbons (PAH) [13-19].

The process of lignin degradation by these fungi is nonspecific and nonstereoselective, which explains why the fungi can mineralize lignin and various organic materials. Several oxidoreductases (including peroxidases, oxidases, and laccases) secreted by ligninolytic fungi are involved in lignin biodegradation. Ligninolytic heme peroxidases are able to break down model compounds representative of the main linkage types in lignin due to their high-redox potential and specialized catalytic mechanisms [1].

Heme-peroxidases form a superfamily of enzymes responsible for numerous biosynthetic and degradative functions. Based on their sequences and catalytic properties, the extracellular fungal lignin peroxidase (LiP), manganese peroxidase (MnP), and versatile peroxidase (VP) belong to class II peroxidases within the superfamily of heme-peroxidases, which includes class I prokaryotic lineage peroxidases and class III classical secretory plant peroxidases. The enzymes oxidize substrates in multi-step electron transfers with the formation of intermediate radical cation, similar to that observed in the well-known horseradish peroxidase. Utilization of H_2O_2 as oxidant, with concomitant production of two water molecules, is general for all peroxidases [20].

The overall catalytic cycle of ligninolytic peroxidases (LiP, MnP, and VP) is similar to that of classical peroxidases, such as horseradish peroxidase, and consists of general three steps:

- the 1st step – O-O bond from H_2O_2 is cleaved by heme Fe^{3+} resulting in the formation of Compound I, containing a ferryl iron porphyrin cation radical $(Fe^{4+}{=}O)^{+}$;
- the 2nd step – the conversion of Compound I to Compound II ($Fe^{4+}{=}O$) causing the oxidation of electron-donor substrates (SH) to cation radicals ($S^{\cdot+}$) in the case of LiP or the oxidation of Mn^{2+} to Mn^{3+} in the case of MnP;

- the 3rd step – the one-electron reduction of SH or Mn^{2+} by Compound II with concomitant conversion of enzymes (Fe^{4+}=O) to their original resting state (Fe^{3+}) [21].

LiP and MnP were initially described in *Phanerochaete chrysosporium* [22-24]. LiP exhibits the high redox potential and ability to oxidize large molecules required for direct oxidation of the lignin polymer and other non-phenolic aromatic substrates. The latter includes veratryl alcohol, the LiP physiological substrate with a mediating role in lignin biodegradation. MnP specifically oxidizes Mn^{2+} to Mn^{3+}, which acts as a diffusible mediator for the oxidation of phenolic substrates and as a starter of lipid peroxidation reactions generating strong oxidizers. Hybrid (versatile) peroxidases sharing the catalytic properties of both LiP and MnP were found in species of *Bjerkandera* and *Pleurotus* [25]. Versatile peroxidase is defined by its capabilities to oxidize the typical substrates of other basidiomycetes peroxidases: (i) Mn^{2+}, the typical MnP substrate (Mn^{3+} being able to oxidize phenols and initiate lipid peroxidation reactions); (ii) veratryl alcohol, the typical LiP substrate; and (iii) simple phenols, which are the substrates of *Coprinopsis cinerea* peroxidase [1].

The catalytic properties required for oxidation of high-redox potential and bulky molecules, are provided by several unique characteristics in the molecular architecture of ligninolytic peroxidases such as the presence of oxidation sites on the protein surface to collect electrons from substrates and deliver them to heme using different pathways.

1. The Structure and Functions of Ligninolytic Enzyme System

1.1. The Structure of Ligninolytic Enzyme System

Wood- and litter-degrading basidiomycetes and some ascomycetes are unique in their ability to degrade the heterogeneous polymer of plant lignin by using the set lignin-modifying enzymes. Ligninolytic fungi produce different combinations of extracellular enzymes [26]. Based on the enzyme production patterns of an array of white rot fungi, A. Hatakka [26] suggested three categories of fungi: (I) LiP-MnP group (e.g. *P. chrysosporium* and *Phlebia radiata*), (II) MnP-laccase group (e.g. *Dichomitus squalens* and *Rigidoporus lignosus*), and (III) LiP-laccase group (e.g. *Phlebia ochraceofulva* and *Junghuhnia separabilima*). Later a new type of ligninolytic peroxidases – versatile peroxidases was described; laccases were found in fungi from "LiP-MnP" group such as *P. chrysosporium.* According these data we suggested the new separation of ligninolytic fungi to six groups (Table 1): (I) LiP-MnP-laccase producing fungi, (II) MnP-laccase producing fungi, (III) MnP-VP-laccase producing fungi, (IV) LiP-VP-MnP producing fungi, (V) LiP-laccase producing fungi. To sixth group we put the fungi with insufficient information, now in these fungi only one ligninolytic enzyme was found. The list given in Table 1 is, of course, not complete and can be expected that the number of ligninolytic fungi and range of degraded compounds will rise further over the next years.

Table 1. The types of enzymatic systems and degradative properties of ligninolytic fungi

Fungi	Group[1]	Family[2]	Compounds degraded	References
I. LiP, MnP, Lac				
Clitocybula dusenii (trace LiP)	WRF	Tricholomataceae	Dyes, humic substances	[30, 31]
Coriolus versicolor	WRF	Coriolaceae	Biphenyl, β-O-4 structure of lignin	[32-34]
Ganoderma spp.	WRF	Ganodermataceae	Remazol Brillian Blue R	[35]
Irpex lacteus	WRF	Steccherinaceae	2,4,6-Trinitrotoluene, pyrene, benz[a]anthracene, endocrine-disrupting compounds	[8, 9, 12, 16, 36, 37]
Nematoloma frowardii (later *Phlebia* sp.)	WRF	Strophariaceae	^{14}C-DHP, humic substances	[31, 38, 39]
Phanerochaete chrysosporium (trace Lac in some conditions)	WRF	Meruliaceae	Nylon, nonylphenol, anthracene, phenanthrene (CO_2), humic substances, benz[a]anthracene, benzo[a]pyrene (CO_2), 2,4-dichlorophenol (CO_2),	[5, 14, 31, 40-45]
Phlebia radiata	WRF	Meruliaceae	2,4,6-Trinitrotoluene	[11, 46]
Phlebia tremellosa	WRF	Meruliaceae	–	[47]
Sclerotium rolfsii	WRF	Atheliaceae	–	[48]
Trametes hirsuta	WRF	Polyporaceae	–	[48]
Trametes versicolor	WRF	Polyporaceae	Nylon, phenanthrene, anthracene, endocrine-disrupting compounds, humic substances	[8, 9, 15, 31, 42, 45, 49]
II. MnP, Lac				
Agaricus bisporus	LDF	Agaricaceae	Lignin	[50-52]

Fungi	Group[1]	Family[2]	Compounds degraded	References
Ceriporiopsis subvermispora	WRF	Coriolaceae	β-5 lignin model dimer	[53]
Cerrena unicolor	WRF	Polyporaceae	–	[54]
Cerrena maxima	WRF	Polyporaceae	–	[55]
Collybia dryophila	LDF	Tricholomataceae	Humic acids (CO_2)	[56]
Coriolus hirsutus	WRF	Coriolaceae	–	[55]
Clitocybula dusenii	WRF	Tricholomataceae	–	[57]
Dichomitus squalens	WRF	Poliporaceae	Azo- and triphenylmethane dyes	[58]
Fomes sclerodermeus	WRF	Poliporaceae	–	[59
Ganoderma lucidum	WRF	Ganodermataceae	Phenanthrene, pyrene	[60-62]]
Gymnopus erythropus (*Collibia erythropus*)	WRF	Tricholomataceae	^{14}C-DHP (CO_2), humic substances (CO_2)	[2]
Hypholoma fasciculare	WRF	Strophariaceae	^{14}C-DHP (CO_2), humic substances (CO_2)	[2, 63]
Kuehneromyces mutabilis	WRF	Strophariaceae	Humic substances	[63]
Lentinus edodes	WRF	Tricholomataceae	–	[64]
Panaeolus sphinctrinus	LDF	Strophariaceae	–	[65]
Phanerochaete flavido-alba	WRF	Meruliaceae	–	[66]
Phlebia brevispora	WRF	Meruliaceae	–	[67]
Physisporinus rivulosus	WRF	Coriolaceae	–	[67]
Rigidoporus lignosus	WRF	Coriolaceae	–	[68]
Stropharia coronilla	LDF	Strophariaceae	Anthracene, bisphenol A benzo[a]pyrene	[10, 14, 69, 70]
Stropharia rugosoannulata	LDF	Strophariaceae	Benzo[a]pyrene, bisphenol A, humic substances	[10, 31, 69]

Table 1. (Continued)

Fungi	Group[1]	Family[2]	Compounds degraded	References
Trametes modesta	WRF	Polyporaceae	–	[48]
III. VP, MnP, Lac				
Panus tigrinus	WRF	Lentinaceae	–	[29, 71]
Pleurotus eryngii	WRF	Lentinaceae	–	[72, 73]
Pleurotus ostreatus	WRF	Lentinaceae	Anthracene, fluorene, pyrene, carbamazepine, benzo[a]pyrene, endocrine disrupting phthalates, humic substances	[7, 9, 13, 19, 42, 63]
Pleurotus pulmonarius	WRF	Lentinaceae	–	[74]
IV. LiP, VP and/or MnP				
Agrocibe sp.	LDF	Bolbitiaceae	Phenanthrene, anthracene, fluorene,	[17]
Bjerkandera adusta	WRF	Coriolaceae	Anthracene, fluorene, humic substances	[13, 31, 75]
Bjerkandera fumosa	WRF	Coriolaceae	–	[27]
Bjerkandera sp. BOS55	WRF	Coriolaceae	Benzo[a]pyrene, humic substances	[14, 31, 76, 77]
Phanerochaete laevis	WRF	Meruliaceae	Benz[a]anthracene	[44]
Phanerochaete sordida	WRF	Meruliaceae	Fluoranthene, azo dyes	[4, 18]
V. LiP, Lac				
Junghuhnia separabilima	WRF	Steccherinaceae	–	[26]
Phlebia ochraceofulva	WRF	Meruliaceae	–	[26]
VI. Insufficient information				
Agaricus blazei (Lac)	LDF	Agaricaceae	–	[78]

Fungi	Group[1]	Family[2]	Compounds degraded	References
Armillaria mellea (Lac)	WRF	Tricholomataceae	–	[79]
Botrytis cinerea (Lac)	WRF[3]	Sclerotiniaceae	–	[80]
Coprinus cinereus (Lac)	LDF	Agaricaceae	–	[82]
Coprinus friesii (Lac, peroxidase)	LDF	Agaricaceae	–	[65]
Coriolopsis rigida (Lac)	WRF	Coriolaceae	–	[83]
Grifola frondosa (Lac)	WRF	Meripilaceae	–	[84]
Lepista irina (VP)	WRF	Tricholomataceae	–	[28]
Marasmius quercophilus (Lac)	LDF	Tricholomataceae	–	[85]
Pleurotus sapidus (Lac)	WRF	Lentinaceae	–	[86]
Pycnoporus cinnabarinus (Lac)	WRF	Polyporaceae	–	[87]
Rigidoporus lignosus (Lac)	WRF	Coriolaceae	–	[88]
Trametes pubescens (Lac)	WRF	Polyporaceae	–	[89]
Trametes trogii (Lac)	WRF	Polyporaceae	–	[68]
Trametes villosa (Lac)	WRF	Polyporaceae	–	[80]
Volvariella volvacea (Lac)	WRF	Pluteaceae	–	[90]

[1]Ecophysiological group: WRF – white rot fungi, LDF – litter-decomposing fungi; [2]67; [3]pathogen; Lac – laccase; DHP – synthetic dehydrogenase polymer lignin

The ligninolytic enzymes were found in fungi belonging to 13 family including Agaricaceae, Atheliaceae, Bolbitiaceae, Coriolaceae, Ganodermataceae, Lentinaceae, Meruliaceae, Pluteaceae, Polyporaceae, Sclerotiniaceae, Steccherinaceae, Strophariaceae, Tricholomataceae (Table 1). The first grop (LiP-MnP-laccase) include generally white rot fungi; LiP was found in the one litter-decomposing fungus – *Agrocibe* sp. VPs are described mainly in *Pleurotus* and *Bjerkandera* species [25, 27] and in some other white rot fungi such as *Lepista irina* [28] and *Panus tigrinus* [29]. In litter-decomposing fungi VP did not found now. MnP and laccase seems to be more widely distributed among ligninolytic fungi than LiP and VP.

The members of first and second groups are the active decomposers of lignin, humic compounds, and different pollutants (Table 1). Some compounds can be mineralized to CO_2 (Table 1). Perhaps, MnP and laccase is the minimal sufficient enzyme complex for appearance of degradative and mineralizing activities.

1.2. The Conditions of Production of Ligninolytic Peroxidases

To date, the studies of lignin-degrading enzymes have been conducted in chemically defined liquid media and different natural substrates. For example, *Nematoloma frowardii* produces MnP as the predominant ligninolytic enzyme with low level of LiP during solid-state fermentation of wheat straw [38] and *Physisporinus rivulosus* T24li produced MnP when it was grown on spruce wood chips [91].

In natural environments, ligninolytic fungi grow on woody substrates. Various reports have indicated that fungal growth in solid-substrate fermentation is different when compared to growth in submerged cultures in laboratory because of the different surface phenomena, moisture content and chemical composition of the substrate [92].

Ligninolytic fungi synthesize lignin-degrading enzymes when grown on media not necessarily requiring natural lignin-containing substrates. Production of ligninolytic peroxidases is a secondary metabolic process that can be triggered by the depletion of nutrients in a medium (such as in *P. chrysosporium*), and it does not depend on the presence of lignin in the cultivation medium [92].

Nitrogen, Carbon, Oxygen

The role of nitrogen, carbon and oxygen in ligninolytic activity varies and that conditions can differ for each individual fungus even within the same species.

The stimulation of production of ligninolytic peroxidases by nitrogen limitation is not general rule. Synthesis of LiP in *P. chrysosporium* may be induced by limiting the sources of nitrogen, carbon and sulphur [92]. The highest MnP levels are produced in nitrogen-limiting conditions during secondary metabolism too and this rule was observed for some white rot fungi, including *P. chrysosporium* [92], *P. sordida* [93], *P. radiata* [46], *P. brevispora* [26], and *C. versicolor* [26]. Under nitrogen limitation conditions *Irpex lacteus* produced three ligninolytic enzymes – laccase, MnP, and LiP, and their formation followed a temporal sequence, with the peak level of laccase activity preceding that of MnP and LiP [37].

However, many exceptions have been described. For example, *Bjerkandera* sp. BOS55 produced LiP in nitrogen-sufficient glucose-peptone medium, whereas no LiP was detectable in N-limited medium. The production of LiP was induced by the peptide-containing components of this medium and also by soy bean protein. Further research showed that the induction of LiP depended on the combination of pH of medium and the type of N source [94]. *Bjerkandera* sp. strain BOS55 is a nitrogen-unregulated white rot fungus which produces high amounts of MnP in the excess of N-nutrients due to increased biomass yield although lower production was also evident in N-limited medium [94, 95]. *C. subvermispora* produced higher levels of MnP when grown under nitrogen- and carbon-sufficient conditions, and *D. squalens* degrades lignin and therefore expresses its lignin-degrading enzymes under nitrogen-sufficint conditions [53, 58, 95]. *I. lacteus* produced MnP under nutrient-sufficient conditions (in malt-extract, glucose, peptone, malt-extract medium) during decolourization of textile industry effluent [97]. MnP was produced by the fungus *P. ostreatus* in liquid medium with peptone as nitrogen source [19]. The obtained results show the importance of media composition in the production of different enzymes within the same fungal species [98].

The effect of amino acids, complex nitrogen sources and vitamin addition on production of ligninolytic enzyme by *T. trogii*, *T. villosa* and *C. versicolor* var. antarcticus, was studied by Levin et al. [99]. Glutamic acid followed by peptone, were the best nitrogen sources for MnP production. Although the response was not uniform, vitamin addition sometimes stimulated ligninolytic enzyme production, but never inhibited it. Thiamine induced MnP production [99]. Collins and Dobson showed that *C. versicolor* produced high level of LiP at high concentration of carbon (10 g/l) [100]. Oxygen also enhanced the production of MnP by *P. sordida.* MnP levels in oxygen-treated cultures exceeded that in air-treated cultures up to 20 times [93].

Detergents

During long time the researches on the extracellular hemeprotein ligninases of *P. chrysosporium* has been hampered by the necessity to produce them in stationary culture. Jager et al. [101] were the first that find that the addition of Tween-80, Tween-20, or 3-[(3-colamidopropyl)dimethylammonio]l-propanesulfonate to the cultures permits development of LiP and MnP activity comparable to that routinely obtained in stationary cultures. The detergent-amended cultures express the entire ligninolytic system, assayed as the complete oxidation of [^{14}C] lignin to $^{14}CO_2$. The detergent effect is evidently not merely in facilitating release of extant enzyme. Development of ligninolytic activity in the agitated cultures, as in stationary cultures, is idiophasic [101].

Other Factors

MnP is also regulated at the level of gene transcription by heat shock and H_2O_2. Some organic acids, such as glycolate, malonate, glucuronate, 2-hydroxybutyrate, added to the medium enhance the production of MnP by the white rot fungus *Bjerkandera* sp. BOS55 [20]. Stimulation of ligninolytic peroxidases by peptone in a nitrogen-rich medium, as found in *Pleurotus,* has also been reported in *Bjerkandera adusta* [19, 102]. Furthermore, production of versatile peroxidase in *Bjerkandera* sp. BOS55 is greatly stimulated by exogenous addition of various physiological organic acids, such as glycolate, glyoxylate, and oxalate [103].

Manganese

The production of ligninolytic enzymes is enhanced significantly if the fungal culture has been induced by lignin or by the compounds having structures similar to those of the products of lignin degradation [92]. Manganese and veratryl alcohol are inducers of these enzymes that have received the most discussion. The requirement for manganese in the culture medium to increase MnP activity is characteristic common among most of the white rot fungi. The first extracellular MnP was purified from *P. chrysosporium*, its expression and production shown to be regulated by the presence of Mn^{2+} in the culture medium. Mn^{2+} controls the mnp gene transcription that is growth and concentration dependent [20]. For example, Mn-peroxidase production by the fungi *P. chrysosporium* and *T. versicolor* was induced not only by C or N starvation but also by Mn^{2+} [104]. Induction of Mn-peroxidase by Mn^{2+} was also found in several fungi, including *P. laevis* [44], *P. sordida* [93], *P. radiata, P. chrysosporium*, *P. subserialis* [105], *Lentinus edodes* [103], *Fomes sclerodermeus* [59] and *Dichomitus squalens* [106].

Litter-decomposing fungi *Agrocibe praecox* and *S. coronilla* produced MnP increasingly in response to Mn^{2+} in the medium. MnP activity was not observed in Mn^{2+}-free cultures of *A. praecox*. *S. coronilla* possesses a partly constitutive MnP which was also detectable in Mn^{2+}-free cultures; although its amount was considerably lower [107].

C. subvermispora grown in the absence of Mn^{2+} showed high levels of mnp transcripts. However, almost no MnP enzyme was detected in the extracellular medium. In the corresponding mycelia, immuno-electron microscopy experiments showed high levels of MnP enzyme within intracellular compartments. Mancilla et al. [96] suggest that in addition to its well-known effect on transcription regulation of mnp genes, manganese influences secretion of MnP to the extracellular medium.

However, the Mn^{2+} repression of a Mn^{2+}-oxidizing peroxidase in *Pleurotus* species has been reported [102]. Mn^{2+} inhibits Mn-peroxidase (versatile peroxidase) production by *P. eryngii, P. ostreatus, P. pulmonarius,* and *P. sajor-caju*, whereas a high level of this enzyme was obtained in a peptone medium [102, 108].

Apparently the induction of ligninolytic enzymes production by Mn^{2+} is common property of the most of the ligninolytic fungi with the exception of *Pleurotus* species.

Aromatic Compounds

Little is known about the role of aromatic compounds in the induction of LiP, MnP and versatile peroxidase, with veratryl alcohol being the most studied [57]. Veratryl alcohol has an importance for synthesis and function of LiP (as its endogenous inducer, stabilizer, and mediator). First this phenomen was found in culture of *P. chrysosporium*. LiP activity is stimulated by incubating cultures with veratryl alcohol, which is a secondary metabolite of this fungus [109]. Increases in LiP activity *C. versicolor* were observed upon supplementation of the medium with veratryl alcohol too [100].

Different strains of *T. versicolor* show different Mn-peroxidase activities when veratryl alcohol or veratric acid is used as an inducer [49]. Mn-peroxidase production is stimulated by veratric acid in *P. radiata* [110]. Gill and Arora [111] showed that Mn-peroxidase production can be promoted by the lignin-degradation products in the white rot fungi such as *D. squalens*, *Irpex flavus*, and *Polyporus sanguineus*. Furthermore, Schlosser et al. [112] showed that when *T. versicolor* is cultivated on wheat straw or beech wood, the water-soluble compounds of lignin or lignocellulose fractions of these solid substrates can induce Mn-peroxidase. Dye effluent increased MnP activity under submerged cultivation of white rot fungus *Clitocybula dusenii* [30]. Sethuraman et al. [113] tested seven benzoic acids, ten cinnamic acids, and five benzaldehyde derivatives for their effects on production of Mn by *C.*

subvermispora FP 90031-sp. and *Cyathus stercoreus* ATCC 36910. Di- and trimethoxylated compounds induced more MnP activities than the corresponding hydroxylated derivatives. Such as syringic acid induced MnP to 536% of that in control for *C. subvermispora*, and both 3,4-dimethoxycinnamic acid and 3,4,5-trimethoxycinnamic acid induced MnP to over 300% of control for *Cyathus stercoreus* [113].

1.3. The Possible Functions of Ligninolytic Enzyme System

As was shown lignin-degrading system played the primary role in the process of degradation of lignin in natural lignocellulosic substrates and a wide range of aromatic xenobiotics, including dyes [3]. Various authors have tried to establish correlations between ligninolytic enzymes and lignin or pollutants degradation.

Lignin biodegradation studies have been carried out mostly using the white rot fungus *P. chrysosporium* which produces multiple isoenzymes of LiP and MnP but does not produce laccase. For a long time LiP was considered as a key enzyme in lignin biodegradation due its ability to catalyze the oxidation of compounds with high redox-potential [22, 23]. Lignin transformation ability may closely reflect the production of ligninolytic enzymes. Both lignin peroxidase and MnP have been demonstrated to perform lignin mineralization and solubilization *in vitro* [2].

LiP produced by some white rot fungi is able to oxidize the non-phenolic substructures of lignin but it seems that LiP is not commonly produced by ligninolytic fungi (Table 1). Evidence for the crucial role of MnP in lignin biodegradation is accumulating, e.g. in depolymerization of lignin and chlorolignin, in demethylation of lignin and delignification and bleaching of pulp, and in mediating initial steps in the degradation of high-molecular mass lignin [26]. MnP from *P. pulmonarius* involved in lignin attack, the addition of Mn^{2+} to the culture increased lignin mineralization by 125% [114]. *N. frowardii* produces MnP as the predominant ligninolytic enzyme during solid-state fermentation of wheat straw but low levels of laccase and LiP also were detected. Synthetic ^{14}C-ring-labelled lignin (^{14}C-DHP) was efficiently degraded during solid-state fermentation and, at last, partial extracellular mineralization of lignin can occurred. In the fermented straw was detected high levels of organic acids (malate, fumarate, oxalate), which rendered MnP effective and therefore made partial direct mineralization of lignin possible [38]. *C. subvermispora* degrades the *β*-5 phenolic lignin model structure. In these experimental conditions, both MnP and laccase were produced whereas LiP was not detected. The depletion of manganese caused a decrease in MnP production. Enzymatic degradation was enhanced by Tween-80 and Tween-20 [53].

As it was mentioned above the white rot fungus *P. ostreatus* produces two types of extracellular peroxidases: MnP and VP. Cohen et al. [115] showed that Mn^{2+} plays a key role in lignin degradation under different nutritional and growth conditions, since it is required for the production of MnP in *P. ostreatus* [115].

In addition, litter-decomposing fungi, which did not produce LiP, are able to degrade and even efficiently mineralize recalcitrant biopolymers such as lignin and humic acids. Therefore, the apparent lack of LiP in litter-decomposing fungi suggests that the fungi may use alternative mechanisms to degrade non-phenolic lignin moieties [70]. MnP is considered to play a key role in the degradation and mineralization of these aromatic polymers, since all fungi showing high mineralization capability produce this peroxidase [70]. Most active species were found within the family *Strophariaceae* (*Agrocibe praecox*, *S. coronilla*, *S. rugosoannulata*). The fungi mineralised around 25% of ^{14}C-DHP to CO_2 and 20% of the lignin were converted to water-soluble fragments. MnP was found to be the predominant ligninolytic enzyme of all fungi in liquid culture and its production was strongly enhanced in the presence of Mn^{2+}. Probably the certain ubiquitous litter-decomposing basidiomycetes possess ligninolytic activity similar to the wood-decaing fungi [116].

Most of the basic research into extracellular nonspecific oxidizing enzymes of ligninolytic fungi has focused on their role in lignin degradation. However, it appears that they also play an important role in various organic material rich environments during the formation, transformation and degradation of humus. In addition, these enzymes are being intensively studied in relation to their ability to degrade a large variety of aromatic organic pollutants. Through evolution, different white rot and litter-decomposing fungi developed different combinations and properties of these enzymes, and they therefore differ in their abilities and mechanisms for the degradation of recalcitrant compounds such as humic substances. The biochemistry of these enzymes as well as the physiology and genetics of their production have been intensively reviewed elsewhere [31].

Strong evidence for the involvement of LiP in degradation of humic substances was found during the incubation of *P. chrysosporium* with brown coal. The crude extract which exhibited high LiP activity was found to depolymerize humic substances. In addition, polymerization of humic substances from coal was detected by using a mutant that was unable to produce LiP but synthesized MnP, suggesting that LiP is the key enzyme in degradation. A mutant that does not secrete either of the enzymes failed to depolymerize brown coal [31]. MnPs from *N. frowardii, C. dusenii* and *Collybia dryophila* was found to degrade and mineralize humic substances originating from brown coal. Moreover, MnP has been suggested as a key enzyme in the degradation and mineralization of humic acids [31]. Correlation was observed between the activity of extracellular peroxidases and humic acid degradation by some white rot and litter-decomposing fungi. Extracellular MnP can be a key enzyme in the process of conversion of humic substances by litter-decomposing fungus *Collibia dryophila* [56].

The non-phenolic subunits that are the most recalcitrant and abundant ones in natural lignin are not susceptible to the oxidation and cleavage by the Mn^{3+}-chelate complexes forming during usual MnP catalysis. MnPs were proposed to initiate the peroxidation of unsaturated fatty acids with production of peroxyl and other lipid radicals. These radicals, in turn, are able to oxidize non-phenolic lignin subunits [70].

Different authors discuss the key role of ligninolytic enzyme system in degradation of wide range of pollutants by white rot and litter-decomposing fungi. For example Golan-Rozen et al. [7] suggest that MnP/VP (in a Mn^{2+}-dependent or independent manner) involved in the oxidation of the pharmaceutical compound carbamazepine in liquid culture of *P. ostreatus*. The white rot fungus IZU-154 significantly degraded nylon-66 membrane under ligninolytic conditions (starvation for carbon or nitrogen) and the process evidently was accelerated by addition of manganese suggesting the participation of MnP. The nylon-degrading activities of *P. chrysosporium* (ATCC 34541) and *T. versicolor* (IFO 7043) were found in this study too [117]. The culture of *P. sordida* showed maximum MnP activity just before the maximum azo dye decolourization was reached. The involvement of MnP-catalyzed lipid peroxidation in dye decolourization by this fungus was proposed [4]. The ligninolytic enzyme system, probably, plays a key role in the initial step of PAH degradation by white rot and litter-decomposing fungi too [for review see 118]. The initial degradation of endocrine-disrupting bisphenol A by *I. lacteus* was accomplished by the activities of MnP and laccase and these enzymes might be involved in bisphenol A degradation [8].

The involvement of ligninolytic peroxidases to utilization of products of degradation of different pollutants was found too. Extracellular peroxidases of *P. radiata* did not reduced 2,4,6-trinitrotoluene but they are involved in the transformation of the first reduction products of parent compound such as 4-hydroxylamino-2,6-dinitrotoluene or 4-amino-2,6-dinitrotoluenent. 4-Hydroxylamino-2,6-dinitrotoluene can be substrate of LiP [11]. By the *in vitro* enzymatic system of *I. lacteus*, which showed a high LiP activity, two hydroxylamino derivatives (2- and 4-hydroxylamino-2,6-dinitrotoluene) were transformed [12].

However, peroxidases can be involved in all steps of xenobiotic degradations. For example, Valli and Gold [5] showed that under secondary metabolic conditions the white rot basidiomycete *P. chrysosporium* mineralizes 2,4-dichlorophenol. The multistep pathway involves the oxidative dechlorination of 2,4-dichlorophenol to yield 1,2,4,5-tetrahydroxybenzene. The intermediate 1,2,4,5-tetrahydroxybenzene is ring cleaved to produce, after subsequent oxidation, malonic acid. In the first step of the pathway, 2,4-dichlorophenol is oxidized to 2-chloro-1,4-benzoquinone by either MnP or LiP. 2-Chloro-1,4-benzoquinone is then reduced to 2-chloro-1,4-hydroquinone, and the latter is methylated to form the lignin peroxidase substrate 2-chloro-1,4-dimethoxybenzene. 2-Chloro-1,4-dimethoxybenzene is oxidized by LiP to generate 2,5-dimethoxy-l,4-benzoquinone, which is reduced to 2,5-dimethoxy-1,4-hydroquinone. 2,5-Dimethoxy-1,4-hydroquinone is oxidized by either peroxidase to generate 2,5-dihydroxy-1,4-benzoquinone which is reduced to form the tetrahydroxy intermediate 1,2,4,5-tetrahydroxybenzene. In this pathway, the substrate is oxidatively dechlorinated by LiP or MnP in a reaction which produces a *p*-quinone. The *p*-quinone intermediate is then recycled by reduction and methylation reactions to regenerate an intermediate which is again a substrate for peroxidase-catalyzed oxidative dechlorination. This unique pathway apparently results in the removal of both chlorine atoms before ring cleavage occurs [5].

Lignin Peroxidase (LiP)

2.1. Molecular Properties and Catalytic Cycle of LiP

LiP (EC 1.11.1.14; diarylpropane: H_2O_2 oxidoreductase) was described over 25 years ago in typical lignin degrading white rot basidiomycete *P. chrysosporium.* The multiple isozymes and isoforms of LiP are secreted by other white rot fungal strains such as *T. veriscolor* [119], *P. radiata* [120], and *Bjerkandera* spp. [76].

The LiP isozymes are glycoproteins of 38-46 kDa, with pI values of 3.2-4.0. LiP has a distinctive property of an unusually low pH optimum near pH 3 [20]. The enzyme contains 1 mol of iron protoporphyrin IX per mole of protein and the crystal structure of *P. chrysosporium* LiP has been described in details [121, 122]. Molecule of LiP contains eight major and eight minor α-helices and a limited β-structure in the proximal domain, has eight Cys residues, all forming disulfide bridges, and two calcium-binding sites, one in each domain, with possible function of maintaining the topology of the active site [122]. The overall reaction of LiP is represented by

1,2-bis(3,4-dimethoxyphenyl)propane-1,3-diol + H_2O_2 ↔ 3,4-dimethoxybenzaldehyde + 1-(3,4-dimethylphenyl)ethane-1,2-diol + H_2O.

LiP is able to direct oxidation of dimeric lignin model compounds such as the non-phenolic β-O-4 aryl-glycerol-aryl ethers which are degraded to veratrylaldehyde and β-ring-derived aromatic derivatives [123, 124]. Oxidative reaction involves initial formation of radical cation via one-electron-oxidation, followed by side-chain cleavage, demethylation, intramolecular addition, and rearrangements. In the mechanism, only the formation of the radical cation is enzyme catalyzed, and subsequent reactions of the substrate are non-enzymatic [20, 122].

The preferred aromatic electron donor for LiP is veratryl alcohol, which is non-phenolic and high-redox potential substrate. During one electron oxidation catalyzing by LiP, veratryl alcohol is oxidized to veratrylaldehyde [125, 126].

LiP possesses two substrate oxidation sites. For the high redox-potential substrates, the oxidation site is localized on the enzyme surface at a catalytic active tryptophan (Trp), where the long-range electron transfer between substrate and heme is started [122, 127]. At this active site the cation radical formation during substrate oxidation is stabilized by acidic surrounding of Trp molecule [128]. It was shown for different LiP and VP cDNA sequences (up to more than 30 sequences available) that this catalytic tryptophan is conserved in all of them and this residue is considered to be essential for the LiP-type activities exhibited by both LiP and VP enzymes [129]. This Trp initiates a long-range electron transfer pathway from the protein surface to the buried heme cofactor [130]. Studies on LiPs and VPs from different fungi suggest a contribution of the protein environment around the catalytic tryptophan to substrate recognition and/or oxidation. An acidic environment lowers the local pH and provides a higher redox potential to the tryptophanyl radical what allow binding of substrate cation radical intermediates acting as enzyme-bound mediators [131]. Dyes are oxidized in the second substrate oxidation site localized at the heme edge, near to surface-exposed Glu [128].

Recently, a new type of LiP was discovered in the fungus *Trametes cervina*. This enzyme exhibits typical LiP-type catalytic properties such as oxidation of high-redox potential substrates such as veratryl alcohol and large molecules for example ferrocytochrome c. LiP from *T. cervina* oxidizes a negatively charged aromatic disulfonate dye as low redox potential ABTS and 1,4-dimethoxybenzene by one-electron oxidation of aromatic ring [131]. However, it lacks the conserved catalytic Trp and contains one tyrosine residue (Tyr 181) at a position where it was never found in other LiPs and VPs [132, 133]. This tyrosine is located near the heme, is accessible to solvent [133] and probably represents an alternative substrate oxidation site [131] like in other different peroxidases (such as poplar peroxidase and heme domain of animal cyclooxygenase) using tyrosine as redox-active residue [134].

2.2. Substrate Specificity of LiP

Catalytic properties of LiP are tested using single-ring, aromatic substrates, including phenolic compounds, such as guaiacol, vanillic acid, and syringic acid, and non-phenolic veratryl alcohol, and dimethyl-phenylenediamine.

Another group of substrates are lignin model dimers acting as synthetic mimics of the common lignin substructures, such as diarylpropane and β-aryl ether dimer [20].

Physiological substrate of LiP is veratryl alcohol and LiP possesses more than two binding sites for this high-redox potential substrate [135]. LiP is relatively nonspecific to its substrates and has been known to oxidize phenolic aromatic substrates and a variety of non-phenolic lignin model compounds as well as a range of organic compounds. Some data on substrate specificity of LiP are summarized in Table 2.

LiP has the unique ability to catalyze oxidative cleavage of carbon-carbon bonds and ether bonds (C-O-C) in high-redox potential non-phenolic aromatic substrates through the formation of a substrate cation radical. LiP catalyzes different reactions such as hydroxylation of benzylic methylene group, oxidation of benzyl alcohols to the corresponding ketones or alcohols, phenol oxidation, and aromatic cleavage of non-phenolic lignin models [136].

LiP can directly interact and oxidize polymeric substrates such as lignin and cytochrome c but this oxidation is dependent on the presence of veratryl alcohol as mediator [135]. LiP oxidizes high-redox potential aromatic compounds but is not able to oxidize phenolic compounds efficiently because of inactivation in the absence of veratryl alcohol. The use of aromatic mediators acting via their cation radicals is required for simple phenols oxidation by LiP [126]. 2-chloro-1,4-dimethoxybenzene was used as mediator during anisyl alcohol oxidation by LiP [137], hydroxybenzotriazole during *in vitro* degradation of kraft pulp [138], and thioanosole during 4-methoxymandelic acid oxidation [139]. It is worthy to note that LiP as strong oxidant (in comparison to other ligninolytic enzymes) does not require the participation of mediators in the oxidation of non-phenolic substrates.

High-molecular weight polymers were produced by LiP from *P. chrysopsorium* in a reaction mixture containing penthachlorophenol and humic acid precursor (ferulic acid) in the presence of detergent and H_2O_2 [41].

2.3. Catalytic Activity LiP towards Lignin and Xenobiotics

LiP is relatively non-specific oxidant with a usually high reduction potential (Table 2). LiP is able to depolymerize synthetic lignin [154], oxidize and transfer PAHs [152], chlorophenols [155], high-redox potential dyes [3, 156] as well as nitro-aromatics and explosives [157]. Podgornik et al. [158] described decolourization more than 40 different dyes by a crude enzyme preparation of LiP and decolourization process was nonspecific and was not dependent on the chromophoric system and the sign and distribution of the charge but slightly dependent on the auxochromic group. LiP from *P. chrysosporium* is able to degrade different synthetic dyes such as methylene blue, Azure B, Congo red, and Crystal violet [159-161] but is not able to degrade Reactive Black 5, high-redox potential dye, without addition of the redox mediator such as veratryl alcohol [156]. LiP catalyzes oxidative decolourization of macromolecules such as melanin [162], macromolecules from a brown coal [163], and is involved in the biobleaching of hardwood kraft pulp by *P. sordida* [164]. It has been demonstrated that LiP, similar like other fungal peroxidases, is able oxidized different chlorophenols such as 2-chlorophenol, 4-chlorophenol, and 2,4-dichlorophenol [5, 161].

Table 2. The main catalytic properties of lignin peroxidise

Substrate	K_M, mM	k_{cat}, s^{-1}	The source of enzyme	Reference
Veratryl alcohol	0.20	–	*P. chrysosporium*	[43]
	–	–		[137]
	0.27	1.85		[139]
	–	–	*Bjerkandera* sp. BOS55	[137]
	–	–	*T. versicolor*	[33]
	0.326 0.234 0.316	17.8 14.9 20.9	*I. lacteus* (three isoenzymes)	[37]
	0.054	–	*Lenzitus betulina*	[140]
Anisyl alcohol	–	–	*P. chrysosporium*	[137]
	–	–	*Bjerkandera* sp. BOS55	
	–	–	*P. chrysosporium*	[141]
H_2O_2	0.017 0.015 0.014	–	*I. lacteus* (three isoenzymes)	[37]
	0.081	–	*Lenzitus betulina*	[140]
Phenol	0.81	–	*P. chrysosporium*	[43, 142]
2,4-Dichlorophenol	–	–	*P. chrysosporium*	[143]
2,4,5-Trichlorophenol	–	–		
2,4,6-Trichlorophenol	–	–		
Pentachlorophenol	–	–	*P. chrysosporium*	[143, 144]
1,4-Dimethoxybenzene	–	–	*P. chrysosporium*	[137]
	–	–	*Bjerkandera* sp. BOS55	
p-Methoxymandelic acid	–	–	*P. chrysosporium*	[141]
3,3-Dimethoxybenzidine	–	–	*P. chrysosporium*	[142]
o-, *p*-, *m*-Phenylenediamine	–	–	*P. chrysosporium*	[142]

Substrate	K_M, mM	k_{cat}, s^{-1}	The source of enzyme	Reference
Aniline	–	–	*P. chrysosporium*	[142]
o-, *m*-Aminophenol	–	–	*P. chrysosporium*	[142]
Catechol	–	–	*P. chrysosporium*	[142]
Pyrogallol	–	–	*P. chrysosporium*	[142]
Guaiacol	–	–	*P. chrysosporium*	[142]
2,6-Dimethoxyphenol	–	–	*P. chrysosporium*	[142]
Methyl orange	–	–	*T. versicolor*	[145]
Eriochrome blue black	–	–	*T. versicolor*	[145]
2-Chloro-1,4-dimethoxybenzene	–	–	*P. chrysosporium*	[137]
	–	–	*Bjerkandera* sp. BOS55	
Tetrachloro-4-methoxyphenol	–	–	*P. chrysosporium*	[137]
	–	–	*Bjerkandera* sp. BOS55	
4-Bromophenol	0.601	41.2	*P. chrysosporium*	[146]
2,4-Dibromophenol	0.387	42.7		
Dimeric lignin model compounds (β-O-4)	–	–	*T. versicolor*	[33]
Dimeric lignin model compounds (β-1)	–	–	*T. versicolor*	[33]
Dimeric lignin model: 1-(3,4-dimethoxy-phenyl)-2-phenoxy-1-ethanol	0.28	0.38	*P. chrysosporium*	[147]
Trimeric lignin model compound (β-O-4)	0.44	0.13	*P. chrysosporium*	[147]
Milled wood lignin	–	–	*P. chrysosporium*	[148]
Benzo[a]pyrene	–	–	*P. chrysosporium*	[148, 149]

Table 2. (Continued)

Substrate	K_M, mM	k_{cat}, s^{-1}	The source of enzyme	Reference
Anthracene	–	–	*P. chrysosporium*	[149-151]
Pyrene	–	–	*P. chrysosporium*	[149, 151, 152]
Fluoranthene	–	–	*P. chrysosporium*	[151]
1-Methylanthracene	–	–	*P. chrysosporium*	[151]
2-Methylanthracene	–	–	*P. chrysosporium*	[149, 151]
9-Methylantracene	–	–	*P. chrysosporium*	[151]
Acenaphthene	–	–	*P. chrysosporium*	[149, 151]
Dibenzothiophene	–	–	*P. chrysosporium*	[151]
Aliphatic halocarbons (reductive dehalogenation)	–	–	*P. chrysosporium*	[153]

The key functional possibility of LiP is that this ligninolytic peroxidase can oxidize aromatic rings that are only moderately activated by electron-donating substituents [129]. Because of this LiP is capable of abstracting an electron from aromatic compounds that carry only two or three ether substituents what is present in lignin polymer. Reactions catalyzed by LiP include Cα-Cβ cleavage of the propyl side chains of lignin and lignin models. Depolymerization of lignin *in vivo* and *in vitro* was demonstrated for LiP from *P. chrysosporium* [154, 165].

Trp 171 residue of LiP is present on the enzyme surface, and is thought to participate in long-range electron transfer from substrates that cannot contact directly with the oxidized heme in LiP active center [128]. For example, LiP is able to oxidize lignin-related substances with a molecular mass greater than 1200 Da [166] or synthetic lignin containing 20 subunits but in the presence of veratryl alcohol cation radical which may act as a diffusible mediator [167]. The addition of veratryl alcohol would circumvent the low efficiency of LiP catalysis on large lignin polymer and might allow LiP to oxidize lignin within plant cell walls which is poorly permeable for enzyme molecule [129]. Additionally veratryl alcohol prevents the enzyme from remaining oxidized for long-time-lasting period of bulky lignin transformation. During this process veratryl alcohol can act as an efficient electron donor to rescue LiP from inactivation by hydrogen peroxide [168].

LiP from *P. chrysosporium* was intensively tested by many authors during effective degradation of different PAHs as benzo[a]pyrene, anthracene, and fluorene [for review see 118]. LiP catalyzes their one-electron oxidation to produce corresponding quinones which can be further metabolized via ring fission [149, 151].

2.4. Influence of pH Values, Temperature and Effectors

The redox potential of all fungal ligninolytic peroxidases is pH-dependent. LiP has very low optimal pH value amounted about 3 and this value reflects the pH required to attain the redox level necessary to oxidize high-redox potential substrates. In the first step of catalysis the native LiP reacts with H_2O_2 and this reaction is independent on pH from pH 2 to 7.5. The rate of the next step, the reduction of Compound I to Compound II during substrate oxidation, is pH-sensitive, with an increasing potential as the pH decreases. The last step of catalytic cycle, the formation of a catalytic inactive form of the enzyme (Compound III), proceeds at pH 3.0, in the presence of excess H_2O_2 and the absence of a reducing substrate. At pH 3 Compound I of LiP stores a high redox potential that enables it to catalyse the oxidation of non-phenolic aromatic substrates not normally associated with other peroxidases. The redox potential decreases linearly as the pH increases, but is not affected by the substrate [20, 122].

LiP-oxidation of veratryl alcohol is influenced by manganese ions, and high Mn^{2+} concentrations markedly repressed veratraldehyde formation [169]. As all peroxidases, LiP is inactivated by the presence of hydrogen peroxide [170] and enzyme could be stabilized by the addition of veratryl alcohol to the reaction mixture. Veratryl alcohol oxidizing by LiP may react with hydrogen peroxide to produce highly reactive dioxygen anion, which then reacts with substrate. Veratryl alcohol, the natural compound producing by fungi, are the main effector of LiP and can enhance the oxidation and the degradation of different compounds such as synthetic lignin, high-redox potential dyes or polycyclic aromatic hydrocarbons.

3. Mn-Dependent Peroxidase (MnP)

3.1. Molecular Properties and Catalytic Cycle of MnP

The first extracellular MnP was purified from *P. chrysosporium*, and later have been reported in large number of fungal species [20, 26].

The molecular mass of MnP ranges from 37 to 62.5 kDa [67, 171], but for the majority of purified enzymes amounts to around 45 kDa [91, 107, 169, 172-174]. Most MnPs characterize acidic isoelectric point – pH 3-4, but in some cases pI values are more neutral around pH 7 [67, 91, 107, 175]. Nevertheless unique MnP from *Trichophyton rubrum* LSK-27 has one of the highest values of pI amounted to 8.2 [176].

Known MnPs are typical fungal glycoproteins created by 10-11 α-helices with the β-structure as a minor component. They contain a heme *b* as a prosthetic group with high-spin iron, pentacoordinated to the four nitrogens in porphyrin core and to the nitrogen of the proximal histidine [177-179]. MnPs structures can be subdivided into three regions. The first one is a distal region containing His, Arg, Asp, and Leu, and this residues are responsible for the reaction of the ferric enzyme with H_2O_2. The second one is a proximal region which is involved in reactivity versus stability of heme and in the stabilization of the Fe^{4+}-state of heme by the proximal His [171]. The third region is responsible for binding of Mn^{2+} ions as the physiological substrate of this enzyme what is highly specific for MnP. Mn^{2+} ion is coordinated to the three carboxylate oxygens of Glu35, Glu39 and Asp179, two water oxygen and a heme propionate oxygen what leads to formation of six-coordinate polyhedron [171, 177, 178]. Furthermore, MnP contains five disulfide bonds and two calcium ions, which are responsible for maintaining the structure of the active site of enzyme molecule [67, 177].

MnP is a one of the fungal ligninolytic enzymes using Mn^{2+} ions as redox cofactor [179]. The catalytic cycle of MnP begins by binding of H_2O_2 to the ferric (Fe^{3+}) enzyme. Then, the bond between oxygen in peroxide is reducing by two electrons, taking away from heme, and one molecule of water and Compound I of MnP are formed. The compound I is a Fe^{4+}-oxoporphyrin-radical complex. The further reduction causes the formation of MnP Compound II, which is Fe^{4+}-oxoporphyrin complex. Mn^{2+} ions are one-electron donors and are oxidized to Mn^{3+} ions by both for of MnP, Compound I and Compound II. The reduction of MnP Compound II leads to formation of native enzyme molecule and to subsequent release of the next water molecule. Mn^{3+} ions are stabilized by organic acids, especially dicarboxylic anions (for instance oxalate or malonate). In this form Mn^{3+} ions are removed from the surface of enzyme and can play a role as the redox-mediators able to oxidize phenolic compounds characterized by lower redox potentials [67, 171, 177, 178].

3.2. Substrate Specificity of MnP

Manganese peroxidase, just like other fungal ligninolytic enzymes, is characterized by low substrate specificity and can oxidize wide range of phenolic and non-phenolic compounds (Table 3). The MnP's substrate can be lignin polymer, specific lignin-derivatives and compounds with structural similarities to those polymers [180].

Obviously, the most specific substrates for manganese peroxidase are Mn^{2+} ions and H_2O_2 with K_M values amounted to 35.2 μM and 6.7 μM, respectively [174]. Other substrates used in measuring of enzyme activity are 2,6-dimethoxyphenol, ABTS and guaiacol but none of these compounds is not specific only for MnP [173]. The rate of these substances oxidation is dependent on the presence of Mn^{2+} [169]. Veratryl alcohol, non-phenolic compound with a relatively high redox potential, is typical substrate for LiP. However, MnP from *P. radiata* is reported to oxidize veratryl alcohol (or other non-phenolic compounds) in the presence of Mn^{2+}, H_2O_2 and a thiol-reductant, for example glutathione. In this reaction the thiyl radical is formed, which acts as an oxidant for other non-phenolic compounds [169]. Some data on substrate specificity of LiP are summarized in Table 3.

3.3. Catalytic Activity towards Lignin and Xenobiotics

MnP was found to catalyze $C\alpha$-$C\beta$ cleavages, $C\alpha$-oxidation and alkyl-aryl cleavages of phenolic syringyl type β-1 lignin structures (Table 3). In these reactions phenoxyl radical intermediates are generated. However, only a small part of lignin polymer substructures consists of phenolic groups, and non-phenolic groups are not susceptible to oxidation by Mn^{3+} chelates. Therefore, two alternative mechanisms have been proposed. In the first one the organic sulphur compounds, such as glutathione, initiate the oxidation of non-phenolic lignin dimers, what led to aryl ether cleavage of the dimers. The second system is based on unsaturated fatty acids generating reactive peroxyl radicals. The MnP-lipid system is able to degrade $C\alpha$-$C\beta$ and β-aryl ether bonds in lignin non-phenolic subuntis [67]. A non-phenolic β-O-4 lignin model compound was partially oxidized in an *in vitro* reaction system developing MnP-dependent lipid peroxidation [70]. Preliminary results showed that crude MnP from *C. subvermispora*, in the presence of Tween-80, was able to oxidize the fully methylated β-5 compound. In contrast, degradation in the presence of Tween-20 was very limited [53].

A non-phenolic lignin model dimer, 1-(4-ethoxy-3-methoxyphenyl)-2-phenoxypropane-1,3-diol, was oxidized by a lipid peroxidation system that consisted of MnP, Mn^{2+}, and unsaturated fatty acid esters. The reaction products included I-(4ethoxy-3-methoxyphenyl)-1-oxo-2-phenoxy-3-hydroxypropane and 1-(4-ethoxy-3-methoxyphenyl)-l-oxo-3-hydroxy propane, indicating that substrate oxidation occurred via benzylic hydrogen abstraction. The peroxidation system depolymerized both exhaustively methylated (non-phenolic) and unmethylated (phenolic) synthetic lignins efficiently [195].

MnP is also directly involved in the degradation of xenobiotic compounds including PAHs, dyes and other organopollutants [3]. PAHs are very hydrophobic compounds with low water solubility and because of it they can easily accumulate on the soil organic matter. The enzymatic oxidation of PAHs frequently requires the addition of glutathione or unsaturated fatty acids (e.g. Tween-80) to the reaction mixture. The presence of these substances allows increasing the degradation of aromatic hydrocarbons. Additionally, surfactants, like Tween-80, can also affect better solvation of the hydrophobic substrates [171, 188]. Eibes et al. [196] have investigated that an efficient degradation system for PAHs requires co-solvents, like acetone, which increased solubility of polyaromatics. They have also reported that MnP from *Bjerkandera* sp. BOS55 is able to degrade anthracene, dibenzothiophene and pyrene in this system [188].

Table 3. The main catalytic properties of Mn-peroxidase

Substrate	K_M, mM	k_{cat}, s^{-1}	Fungus	Reference
H_2O_2	0.016	121.0	*Panus tigrinus* CBS 577.79	[181]
	0.011 0.007	415.0 84.0	*Fomes sclerodermeus*	[59]
	0.032	14.0	*Fomes durissimus* MTCC-1173	[182]
	0.039	–	*P. chrysosporium*	[183]
	0.002	–	*A. praecox*	[107]
	0.002	–	*S. coronilla*	[107]
	0.007	–	*Schizophyllum* sp.	[184]
$MnSO_4$	0.124	242.4	*Panus tigrinus* CBS 577.79	[181]
	0.030 0.027	581.0 285.0	*Fomes sclerodermeus*	[59]
	0.045	161.0	IZU-154	[185]
	–	–	*I. lacteus*	[97]
	0.017	–	*A. praecox*	[107]
	0.012	–	*S. coronilla*	[107]
	0.070	–	*T. versicolor*	[186]
	0.059	22.4	*Fomes durissimus* MTCC-1173	[182]
	0.041	211.0	*P. chrysosporium*	[183]
	0.035	–	*Schizophyllum* sp.	[184]
ABTS	0.196	20.2	*Panus tigrinus* CBS 577.79	[181]
	0.667	–	*A. praecox*	[107]
	0.048	–	*Schizophyllum* sp.	[184]
2,6-Dimethoxyphenol	0.076	26.3	*Panus tigrinus* CBS 577.79	[181]
	0.029 0.084	257 51	*Fomes sclerodermeus*	[59]
	0.123	–	*Schizophyllum* sp.	[184]
Guaiacol	0.013	–	*Schizophyllum* sp.	[184]
	0.213	8.9	*Panus tigrinus* CBS 577.79	[181]
Pyrogallol	0.293	21.0	*Panus tigrinus* CBS 577.79	[181]
Veratryl alcohol	0.560	0.45	*Panus tigrinus* CBS 577.79	[181]
	–	–	*P. radiata*	[169]

Substrate	K_M, mM	k_{cat}, s^{-1}	Fungus	Reference
Anthracene	–	–	*Bjerkandera* sp. BOS55, *Anthracophyllum discolor*, *I. lacteus*, *P. chrysosporium*, *N. frowardii*, *S. coronilla*	[14, 107, 150, 187, 188, 189, 190]
Pyrene	–	–	*Anthracophyllum discolor*, *I. lacteus*, *N. frowardii*	[187, 189, 190]
Fluoranthene	–	–	*Anthracophyllum discolor*, *I. lacteus*, *N. frowardii*	[187, 189, 190]
Phenanthrene	–	–	*Anthracophyllum discolor*, *I. lacteus*, *P. chrysosporium*, *N. frowardii*	[187, 189, 190, 191]
Fluorene	–	–	*P. chrysosporium*	[44]
Chrysene	–	–	*N. frowardii*	[187]
Benz[a]anthracene	–	–	*N. frowardii*	[187]
Benzo[a]pyrene	–	–	*N. frowardii, S. coronilla*	[14, 107, 187]
Benzo[b]fluoranthene	–	–	*N. frowardii*	[187]
Phenolic lignin model dimer	–	–	*P. chrysosporium*	[192]
^{14}C-DHP	–	–	*N. frowardii* (later *Phlebia* sp.)	[38,39]
Non-phenolic β-O-4 lignin model	–	–	*S. coronilla*	[70]
Humic acid (depolimerization)	–	–	*Fomes durissimus* MTCC-1173	[182]
KBr, KI (haloperoxidase activity)	–	–	*Fomes durissimus* MTCC-1173	[182]
Oxalate, glyoxylate	–	–	*Ceriporiopsis subvermispora*	[193]
Nylon	–	–	IZU-154	[117]
Tryphenylmethane dyes	–	–	*I. lacteus*	[97]
Azo dyes	–	–	*I. lacteus*, *Bjerkandera* BOS55, *P. sordida*, *Schizophyllum* sp.	[4, 77, 97, 184]
Polymeric dyes	–	–	*I. lacteus*	[97]
Anthraquinone dyes	–	–	*I. lacteus*	[97]
Heterocyclic dyes	–	–	*I. lacteus*	[97]
Lipid peroxidation	–	–	*S. coronilla*	[97]
Bisphenol A	–	–	*S. coronilla*	[10]
β-5 lignin model compound	–	–	*Ceriporiopsis subvermispora*	[53]
2-Amino-4,6-dinitrotoluene	–	–	*N. frowardii* (CO_2)	[194]

The stimulatory effect of glutathione was demonstrated by Gunther et al. [197], who reported that the addition of mediating agents, such as reduced glutathione, increased the oxidative strength of MnP from *N. frowardii*; as a result anthracene was completely reduced and 60% of pyrene was degraded.

Fungal MnPs have been shown to possess potential in degrading dyes (Table 3) because they are highly oxidative and substrate-nonspecific [174]. The MnP from *Stereum ostrea* was able to decolourize Rose Bengal, Remazol Brilliant Violet, Remazol blac-5, Remazol blue-19 and Remazol orange-16 within a range 70-80% during 16 hours of incubation [180]. The decolourization process was observed also by MnP from *Ischnoderma resinosum, I. lacteus, P. chrysosporium, T. versicolor, P. ostreatus* [198-200]. MnPs can also oxidize some non-phenolic pollutants such as aflatoxin B1 (AFB1), which is difuranocoumarin-derivative mycotoxin with carcinogenic, teratogenic and hepatotoxic properties. MnP from *P. sordida* YK-624 is able to detoxification of AFB1 and removing of its toxicity [201]. MnP from white rot fungus IZU-154 degraded nylon by unknown mechanism [45].

High-molecular weight polymers were produced by LiP or MnP from *P. chrysopsorium* in a reaction mixture containing penthachlorophenol and humic acid precursor (ferulic acid) in the presence of detergent and H_2O_2 [41].

The first description of direct enzymatic mineralization of PAHs by MnPs indicates their important role in the oxidation of PAHs by wood-decaying fungi was showed by Sack et al. [187]. Degradation of PAHs by crude MnP preparation of *N. frowardii* was demonstrated for a mixture of eight different compounds (phenanthrene, anthracene, pyrene, fluoranthene, chrysene, benz[a]anthracene, benzo[a]pyrene, and benzo[b]fluoranthene) and for five individual PAHs, including phenanthrene, anthracene, pyrene, fluoranthene, and benz[a]anthracene. The oxidation of PAHs was enhanced by the addition of glutathione, a mediator substance, which is able to form reactive thiyl radicals. Glutathione-mediated MnP was capable of mineralizing ^{14}C-pyrene, ^{14}C-anthracene, ^{14}C-benzo[a]pyrene, ^{14}C-benz[a]anthracene, and ^{14}C-phenanthrene [187].

Later the direct involvement of manganese peroxidase (MnP) in the mineralization of natural and xenobiotic compounds was evaluated. Crude MnP from the agaric white rot fungus *N. frowardii* oxidized *in vitro* several ^{14}C-PAHs, including pyrene and benzo[a]pyrene, leading to the formation of significant amounts of $^{14}CO_2$ ("enzymatic combustion"); mineralization increased 3- to 10-fold when reduced glutathione was present in the reaction solution. The glutathione effect was attributed to the transient formation of particularly reactive thiyl radicals [197].

A broad spectrum of aromatic substances was partially mineralized by the MnP system of the white rot fungus *N. frowardii*. The cell-free MnP system partially converted several aromatic compounds, including ^{14}C-pentachlorophenol, ^{14}C-catechol, ^{14}C-tyrosine, ^{14}C-tryptophan, ^{14}C-pyrene, and ^{14}C-2-amino-4,6-dinitrotoluene to $^{14}CO_2$. Mineralization was dependent on the ratio of MnP activity to concentration of reduced glutathione. The formation of highly polar products was also observed with ^{14}C-2-amino-4,6-dinitrotoluene and ^{14}C-pentachlorophenol; these products were probably low-molecular-weight carboxylic acids.

Among the aliphatic compounds tested, glyoxalate was mineralized to the greatest extent. Eighty-six percent of the $^{14}COOH$-glyoxalate and 9% of the ^{14}CHO-glyoxalate were converted to $^{14}CO_2$, indicating that decarboxylation reactions may be the final step in MnP-catalyzed mineralization. The extracellular enzymatic combustion catalyzed by MnP could represent an important pathway for the formation of carbon dioxide from recalcitrant xenobiotic compounds and may also have general significance in the overall biodegradation of resistant natural macromolecules, such as lignins and humic substances [202].

The mineralization of benzo[a]pyrene and anthracene by MnP from the litter-decomposing fungus *S. coronilla* was also shown by Steffen et al. [14]. The oxidation of benzo[a]pyrene leads to benzo[a]pyrene-1,6-quinone as a temporary intermediate [14].

MnP from *N. frowardii* mineralized 2-amino-4,6-dinitrotoluene to CO_2 in the presence of reduced glutathione as secondary thiol mediator [194].

Linoleic acid, the predominant unsaturated fatty acid in the lipids of wood-rotting fungi, was oxidized by MnP from the white rot fungus *P. radiata* through a peroxidation mechanism. The peroxidation was markedly stimulated by hydrogen peroxide [203]. MnP of the litter-decomposing basidiomycete *S. coronilla* showed prooxidant activity promoting the peroxidation of linoleic acid in the presence of acetate and Mn^{2+} ions, even without the addition of hydrogen peroxide [70].

MnP from *P. radiata* and *N. frowardii* was found to decompose malonate oxidatively in the absence of H_2O_2 in a reaction system consisting of the enzyme, sodium malonate and Mn^{2+}. The enzymatic oxidation resulted in a substantial decrease in malonate concentration and the formation of CO_2, oxalate, glyoxylate and formate. Simultaneously with the decomposition of malonate, Mn^{2+} was oxidized to Mn^{3+} leading to high transient concentrations of the latter [204].

3.4. Influence of pH Values, Temperature and Effectors

The optimal values of pH and temperature for MnP activity are similar among different fungal strains and are from 4.5 to 6.8 for pH values and from 35°C to 40°C for temperature [91, 169, 173, 174, 176, 180]. Some MnPs show high stability in wide pH range (4.0 – 7.0) as for example the enzymes isolated from *Lentinula edodes* and *Schizophyllum* sp. F17 [173, 174]. Moreover MnP from *Aspergillus terreus* LD-1 is called alkaline manganese peroxidase and have got unique property because of its stability in the pH values from 11.0 to 12.5, with optimal activity in the pH 12.5 [205]. MnP is sensitive to high temperatures and it can be easily inactivated by temperature in short time. However, some fungal strains produce thermostable enzymes such as MnP from *Lenzites betulinus*, which retaines about 60% of the initial activity after one-hour-long treatment in 60°C [175] and MnP from *I. lacteus*, very active in temperature in the range from 50°C to 60°C [189].

Mn^{2+} ions and hydrogen peroxide are natural substrates for MnP and the concentrations of these compounds can affect its enzymatic activity. It has been demonstrated, that ABTS, guaiacol, vanillalacetone and 2,6-dimethoxyphenol were faster oxidized in the presence of Mn^{2+} than in the absence of Mn^{2+} [169, 176]. In some cases high concentration of H_2O_2 may cause gradually inactivation of enzyme. Bermek et al. [176] have investigated that H_2O_2 concentration 1000-fold higher than the MnP's concentration led to 45% decrease of enzymatic activity during the first 15 minutes of reaction. Nevertheless, MnP from *Lenzites betulinus* shows high tolerance to hydrogen peroxide and retaining 60% of residual activity in reaction with 10 mM H_2O_2 [175]. Anethole, eugenol, ethanol, SDS, NaN_3, EDTA, thiourea, Hg^{2+}, Ag^{+}, Pb^{2+}, Fe^{2+}, Cu^{2+} are factors which partially or substantially cause the inhibition of MnP activity whereas Zn^{2+} and Ca^{2+} ions or carboxylic acids such as oxalate, maleate, and fumarate highly stimulate the activity of fungal MnP [91, 173, 174, 180, 205].

4. Versatile Peroxidase (VP)

4.1. Molecular Properties and Catalytic Cycle

Versatile peroxidases (VP) (EC 1.11.1.16) are described mainly in *Pleurotus* and *Bjerkandera* species [20, 27, 206] and in some other fungi such as *Lepista irina* [28] and *Panus tigrinus* [29]. The ability of versatile peroxidase to oxidize both Mn^{2+} and aromatic compounds was first reported in *Pleurotus eryngii* [102]. To the present day, most research on versatile peroxidases has been performed with the enzymes of the white rot fungi *Bjerkandera* sp. strain BOS55 [103], *Bjerkandera* sp. (B33/3) [207], *B. adusta* [208], *B. fumosa* [27, 206], *P. ostreatus*, *P. eryngii*, and *P. pulmonarius* [73].

As ligninolytic peroxidase VP combines both MnP and LiP activities [209] and because of these unique properties is capable of degrading substrates specific for MnP and LiP and is called a hybrid enzyme. Fungal VP possesses different catalytic sites in enzyme molecule which are high related to those in MnP and LiP [210] and is responsible for the oxidation of low- and high-redox-potential compounds.

Additionally VP can oxidize a wide range of substrates because of its high redox potential as well [1]. It is worth noting that VP's structure is more relating to the structure of LiP (58-60%) than to the structure of MnP (55%) [210].

Fungal VP can catalyse various substrates by the two reaction strategies, i.e., via direct oxidation and via Mn^{3+} ions acting as diffusible oxidizers, which can penetrate substrates that are not accessible to enzymes due to steric hindrance. These two mechanisms can be applied alternatively or sequentially with the same enzyme molecule [27].

Generally the VP molecule is formed by 12 helices and heme domain in the middle of the structure and two narrow channels connecting this domain with protein surface [211]. The main channel is larger and is conserved in all peroxidases. Through this channel heme can be reached by hydrogen peroxide to react with Fe^{3+} and two-electron activated Compound I is forming (the 1st step). The second channel connects directly to heme propionate and is used by Mn^{2+} to reach and by Mn^{3+} to exit, similar like in MnP [130]. A Mn^{2+} binding site, containing three carboxy residues from amino acids (Glu36, Glu40 and Asp175), typical of MnP, is found near heme internal propionate [1]. In VP molecule are present residues, typical of LiP, involving in enzyme interaction with veratryl alcohol. The VP from white rot fungus *P. eryngii* contains 12 helices, 4 disulfide bonds, 2 structural Ca^{2+} sites, Mn^{2+} binding site, and heme pocket containing histidine residues like distal His47 and proximal His169. The Mn^{2+} ions are coordinated by the heme proprionate D and by carboxylates of Glu36, Glu40, and Asp175 [20].

Besides of the Mn-biding side, there is also catalytic Trp 164 residue (for *P. eryngii*), an analogus of Trp 171, in LiP [129]. Both Trp 164 and Trp 171 are important for long-range electron transfer pathway and for the oxidation of high-redox-potential compounds, for example veratryl alcohol [211]. The slightly different structure presents the molecule of VP from *B. adusta* where heme iron is connected by domain of H182 and distal domain H53 [21].

4.2. Substrate Specify of VP

The most important and characteristic substrate of VP is Mn^{2+} which can be oxidized to Mn^{3+}, but also phenols (including phenolic units of lignin) and others low-redox potential compounds, typical substrates of MnP (Table 4). Second often tested substrate for VP is veratryl alcohol, which together with other non-phenolic lignin compounds, polycyclic aromatic hydrocarbons, dyes and high-redox-potential compounds, is substrate specific for LiP-site of VP (Table 4) [212, 213]. Mentioned compounds can be oxidized by VP directly, while LiP requires the presence of veratryl alcohol to carry this process [211]. In the case of VP, Mn^{2+} ions act as mediator during oxidation of different compounds and in the most cases, the activity of VP is higher in presence of manganese ions compared to other aromatic effectors [212]. VP is able to Mn^{2+}-independent oxidation of some substrates like phenol monomers and amines [213]. Fungal VPs can also transform different recalcitrant aromatic compounds such as 2,4-dichlorophenol in the absence of Mn^{2+} [108], dichlorophen, bromoxynil, and pentachloroquinone both with and without Mn^{2+} [214]. The versatile peroxidase from species of *Bjerkandera* genus can directly transform lignin polymer, even in the absence of external mediators [207], some pharmaceutical xenobiotics [215], industrial dyes [27, 210], and three- and four-ring polycyclic aromatic hydrocarbons [216, 217].

VP has high affinity for Mn^{2+}, ferulic acid, *a*-naphthol, and different hydroquinones, and dyes, but the affinities for veratryl alcohol and substituted phenols are lower. It shows also high affinity for H_2O_2 (Table 4). VP has the highest V_{max} on Mn^{2+}, moderate activities on different aromatic substrates and the lowest V_{max} on guaiacol and other dyes. The VP affinities for H_2O_2 and Mn^{2+} are higher than obtained for *P. chrysosporium* peroxidases, but veratryl alcohol affinity is lower than found for LiP. The maximal activities on Mn^{2+} and veratryl alcohol are also lower than obtained for MnP and LiP respectively [73].

During VP-mediated oxidation of 2,7-diaminofluorene [216], 2,6-dimethoxyphenol, ABTS, hydroquinone [218], and methoxyhydroquinone [219] both with and without Mn^{2+}, biphasic kinetics with apparent saturation in both micromolar and millimolar ranges were obtained (Table 4).

Formation of H_2O_2 during the oxidation of hydroquinones by VP from *P. eryngii* was investigated by Gomez-Toribio et al. [222]. From several possible buffers, tartrate buffer was selected to study the oxidation of hydroquinones as it did not support the Mn^{2+}-mediated activity of VP in the absence of exogenous H_2O_2. In the absence of Mn^{2+}, efficient hydroquinone oxidation by VP was dependent on exogenous H_2O_2. Under these conditions, semiquinone radicals produced by VP autoxidized to a certain extent producing superoxide anion radical that spontaneously dismutated to H_2O_2 and O_2. The use of this peroxide by VP produced quinone in an amount greater than equimolar to the initial H_2O_2. In the presence of Mn^{2+}, exogenous H_2O_2 was not required for complete oxidation of hydroquinone by VP. Reaction blanks lacking VP revealed H_2O_2 production due to a slow conversion of hydroquinone into semiquinone radicals (probably via autooxidation catalysed by trace amounts of free metal ions), followed by O_2 production through semiquinone autooxidation and O_2 reduction by Mn^{2+}. This peroxide was used by VP to oxidize hydroquinone that was mainly carried out through Mn^{2+} oxidation [222].

Table 4. The main catalytic properties of versatile peroxidise

Substrate	K_M, mM	k_{cat}, s^{-1}	Fungus	Reference
H_2O_2	15.9-25.5	–	*Bjerkandera* sp.[1]	[212]
	0.009 0.010	–	*P. eryngii*[1]	[219]
	0.015	–	*P. ostreatus*	[98]
	0.005	225	*B. adusta*	[220]
	0.031	75	*Bjerkandera* sp. BOS55	[103]
2,6-Dimethoxyphenol	0.078 10.50	5.6 29.8	*P. eryngii*[2]	[218]
	–	–	*Panus tigrinus* 8/18	[29]
	26.6-99.6	–	*Bjerkandera* sp.[1]	[212]
	0.345	10.6	*B. fumosa*	[216]
	0.145	70	*B. adusta*	[220]
	0.041	2.3	*Bjerkandera* sp. BOS55	[103]
2,6-Dimethoxyphenol+$MnSO_4$	20.1-37.9	–	*Bjerkandera* sp.[1]	[212]
	0.063	30.0	*B. fumosa*	[216]
	0.011	–	*P. ostreatus*	[98]
	0.004	320	*B. adusta*	[220]
$MnSO_4$	0.181	275	*P. eryngii*[1]	[1]
	0.048 0.200	79 78		[219]
	14.3-45.1	–	*Bjerkandera* sp.[1]	[212]
	0.68	3.9	*B. fumosa*	[216]
	0.10	–	*P. ostreatus*	[98]
	0.017	150	*B. adusta*	[220]
	0.051	59	*Bjerkandera* sp. BOS55	[103]
ABTS	0.003 1.090	8.1 186	*P. eryngii*[2]	[218]
	–	–	*Panus tigrinus* 8/18	[29]
	0.48	2.77	*B. fumosa*	[216]
	0.013	139	*B. adusta*	[220]
	0.037	1.3	*Bjerkandera* sp. BOS55	[103]

Substrate	K_M, mM	k_{cat}, s^{-1}	Fungus	Reference
ABTS + Mn^{2+}	0.86	6.2	*B. fumosa*	[216]
	0.015	120	*B. adusta*	[220]
Hydroquinone	0.016 0.716	10.3 35.1	*P. eryngii*[2]	[218]
	–	–	*Panus tigrinus* 8/18	[29]
Catechol	5.040	185	*P. eryngii*	[218]
NADH (oxidase reaction)	0.040	–	*P. ostreatus*	[98]
Methoxyhydroquinone	0.017 0.890 2.530	4 19 19	*P. eryngii*[1,2]	[219]
p-Methoxybenzene	2.400	4	*P. eryngii*	[219]
Syringol	0.20 1.00	6 3	*P. eryngii*[1]	[219]
Guaiacol	11.100	22.7	*P. eryngii*	[218]
	–	–	*Panus tigrinus* 8/18	[29]
	4.96	14.5	*B. adusta*	[220]
Guaiacol + Mn^{2+}	0.222	31.7	*B. adusta*	[220]
Veratryl alcohol	4.130	9.5	*P. eryngii*	[1]
	1.287	–	*Bjerkandera* sp.	[212]
	3.50	4	*P. eryngii*	[219]
	3.04	7.08	*B. fumosa*	[216]
	0.116	2.8	*Bjerkandera* sp. BOS55	[103]
	4.100	–	*P. ostreatus*	[98]
	5.300	5.3	*B. adusta*	[220]
Veratryl alcohol + Mn^{2+}	0.53	26.8	*B. fumosa*	[216]
	0.432	39.5	*B. adusta*	[220]
Reactive Black 5	0.003	5.5	*P. eryngii*	[1]
	0.002	5		[219]
p-Phenylendiamine	–	–	*Panus tigrinus* 8/18	[29]
o-Dianisidine	–	–	*Panus tigrinus* 8/18	[29]
2,4,6-Trichlorophenol	–	–	*Panus tigrinus* 8/18	[29]
Syringaldazine	0.066	0.22	*B. fumosa*	[216]
Syringaldazine + Mn^{2+}	0.035	6.1	*B. fumosa*	[216]
2,7-Diaminofluorene	0.009 2.3	0.19 0.45	*B. fumosa*[2]	[216]

Table 4. (Continued)

Substrate	K_M, mM	k_{cat}, s^{-1}	Fungus	Reference
2,7-Diaminofluorene + Mn^{2+}	0.038 0.27	0.83 1.28	*B. fumosa*[2]	[216]
Reactive Violet 5	0.011	–	*B. adusta*	[221]
	0.016[1] 0.012	–	*P. eryngii*	[221]
	0.047	16.9	*B. adusta*	[210]
Reactive Black 5	0.006	–	*B. adusta*	[221]
	0.022	4.8	*B. adusta*	[210]
Reactive Blue 38	0.006	–	*B. adusta*	[221]
	0.008[1] 0.007	–	*P. eryngii*	[221]
	0.089	19.8	*B. adusta*	[210]
Rcactive blue 72	0.027	10.0	*B. adusta*	[210]
Anthracene	–	–	*B. fumosa*	[216]
	–	–	*B. adusta*	[217]
Phenanthrene	–	–	*B. fumosa*	[216]
Fluorene	–	–	*B. fumosa*	[216]
Pyrene	–	–	*B. fumosa*	[216]
	–	–	*B. adusta*	[217]
Chrysene	–	–	*B. fumosa*	[216]
Fluoranthene	–	–	*B. fumosa*	[216]
Benzo[a]pyrene	–	–	*P. eryngii*	[108]
	–	–	*B. adusta*	[217]

[1]Different isoenzymes; [2]Biphasic kinetic.

4.3. Catalytic Activity towards Lignin and Xenobiotics

The unique catalytic properties of versatile peroxidases, including the oxidation of high- and low-redox potential substrates, have made them attractive for biotechnological processes, for example for the oxidation of dyes, polycyclic aromatic hydrocarbons, halogenated pesticides, like dichlorophen, bromoxynil and penthachlorophenol [20, 214]. As well as other lignin degrading enzymes, VP makes an alternative to the use of chemicals in paper industry [223]. VP from *Bjerkandera* sp. strain BOS55 was tested in a eucalyptus kraft pulp biobleaching [224]. This enzyme reacts with soluble lignin fragments without mediators, and the products of this transformation were lignin depolymerized fractions and low-molecular weight compounds [207]. Recently studies focus on enzymatic activation of lignin for improving ligninocellulosic products adhesion and very effective catalyst of this process was two-stage VP/Mn^{3+}-malonate system. During that process VP produces Mn^{3+} ions which next oxidize the target compound by Mn^{3+} reactive species [223]. The versatile peroxidase from *Bjerkandera* sp. (B33/3) can directly transform lignin polymer, even in the absence of external mediators [207].

A small part of research on versatile peroxidases has been addressed to the oxidation of recalcitrant aromatic compounds such as xenobiotics. For example, the enzyme from the white rot fungus *P. eryngii* can degrade 2,4-dichlorophenol in the absence of Mn^{2+} [108]. The versatile peroxidase of *B. adusta* UAMH 8258 transforms dichlorophen, bromoxynil, and pentachloroquinone both with and without Mn^{2+}, and the dehalogenation reactions mediated by this enzyme were also found [214]. Purified VP from the white rot fungus *B. adusta* UAMH 8258 was used to study the transformation of several pesticides, including some as highly halogenated as the wood preservative pentachlorophenol. It was found that dichlorophen, bromoxynil and pentachlorophenol were transformed by VP in the presence and in the absence of Mn^{2+}. The activity was higher in the absence of Mn^{2+} at pH 3 than in the presence of Mn^{2+} at pH 4. Analysis of reaction products showed the presence of 2,3,5,6-tetrachloroquinone among the products from pentachlorophenol oxidation, while the main product from dichlorophen was 4-chlorophenol-2,2-methylenequinone. Several polymers were obtained from the peroxidase oxidation of bromoxynil. In all cases, dehalogenation reactions mediated by the versatile peroxidase were found [214].

The degradation of some pharmaceutical xenobiotics [215] and industrial dyes [210] by VP from *B. adusta* was described. The first report on PAH oxidation by the versatile peroxidase of white rot fungi appeared in 2003 and author described that this enzyme catalyzes oxidation of anthracene and its methyl derivatives, pyrene, and benzo[a]pyrene [217]. Later, the oxidation of benzo[a]pyrene was also observed in reactions with VP from *P. eryngii* [108] and anthracene, phenanthrene, fluorene, pyrene, chrysene, and fluoranthene in rections with VP from *B. fumosa* [216]. 9,10-Anthraquinone, 9,10-phenanthrenequinone, and 9-fluorenone as a product of oxidation of anthracene, phenanthrene, and fluorene by VP from *B. fumosa* were identified respectively [216].

Unlike general peroxidases, VP from *P. ostreatus* has a unique property of direct oxidization of high-molecular-weight compounds, such as Poly R-478 and RNase A. The obtained data demonstrated that, in addition to veratryl alcohol, the high-molecular-weight substrates are directly oxidized by VP [225].

4.4. Influence of pH Values, Temperature and Effectors

Recent studies revealed that optimal VP temperature and pH values from different fungal species are basically the same. Nevertheless range of temperature and pH in which VP shows catalytic activity are very wide and this peroxidase is more thermostable than pH-dependent. VP from *Cerrena* sp. was stable during 3 hours of incubation in 30°C but only during 1 hour after incubation in pH 7.5. These studies were carried out with using pH and temperature values, in which VP have shown more than 50% of its original activity but the optimal values were 4.6 for pH 4.6 and 50°C temperature [226]. The VP from *P. eryngii* was active in the temperatures from 30°C to 80°C and the optimal temperature was 50°C. Regarding pH dependencies this VP showed stability in pH range of 3.6-4.6 but the highest stability was in pH 4 during 24 h of incubation at 4°C [209].

In the case of VP as a hybrid enzyme, temperature and pH optimization can be consider not only for entire enzyme molecule, but also for each catalytic MnP and LiP site. This kind of research was conducted on VP from *B. adusta* in absence or presence of Mn^{2+}, for LiP and MnP site respectively. LiP which is Mn^{2+}-independent and in the absence of manganese ions showed the highest activity in pH 3.0-3.5 while the optimal pH for MnP (Mn^{2+}-dependent) was 4.5 [21]. The higher optimal pH value for MnP site compared to LiP site in VP from *B. adusta* is likely to be caused by the three acidic residues and the heme-propionate dissociating at a higher pH values and binding Mn^{2+} [21, 219]. What is interesting in some other tribe of *B. adusta* (UAMH 8258) found that during reactions of VP with different substrates such as pesticides, polycyclic aromatic hydrocarbons, phenolic and non-phenolic compounds, the best pH values were pH 3.0 for MnP site and pH 4.0 for LiP site [210, 220]. Wang et al. [220] found that peroxidase from *B. adusta*, UAMH 8258 exhibited both Mn^{2+}-dependent and Mn^{2+}-independent activity and the pH optimum for the former was pH 5.0 for five tested substrates (2,6-dimethoxyphenol, veratryl alcohol, hydroquinone, 4-aminophenol, and ABTS), while the optimum pH for the latter was pH 3.0 for 2,6-dimethoxyphenol, ABTS and veratryl alcohol [220].

Besides substrate promiscuity the VP activity towards certain compounds may be additionally modified by some sort of effectors, for example Mn^{2+}, which is an effector often replaced in literature [21]. During oxidation of some compounds by VP, manganese ions may act as a substrate and as an inhibitor as well. VP was unable to oxidize compounds like veratryl alcohol, Reactive Black 5 and 2,4-dichlorophenol at concentrations greater than 0.1 M. On the other hand, lack of Mn^{2+} may also cause inhibition in substrate oxidation, for example *p*-aminobenzoic acid. In this case, VP from *B. fumosa*, could not oxidize this substrate without Mn^{2+} presence because of lack diffusible chelated Mn^{3+} acting as the oxidizing agent [216]. However, there are compounds which can be oxidized by this enzyme with and without Mn^{2+}, for example ABTS, 2,7-diaminofluorene, syringaldazine, veratryl alcohol, anthracene [216]. But in some cases, like oxidation of PAHs, manganese ions are necessary factor for this process or significantly improve the oxidation of various kinds of substrates like hydroquinone, 2,6-dichlorophenol, hydrocaffeic acid *N*-ethyl-*N*-(2-hydroxy-3-sulfopropyl)-3-methylaniline and 2,6-dimetoxyphenol [226].

In comparison to LiP, versatile peroxidase is able to oxidize veratryl alcohol, but it's affinity to that substrate is lower than LiP [216]. There are some compounds for example dyes which are oxidized by VP more effectively in the presence of veratryl alcohol. For example VP from *Cerrena* sp. is able to decolourize Acid Red 52 and Phenol Red in the presence of veratryl alcohol [226]. Furthermore VP from *B. adusta,* called as "dye-peroxidase", is more effective during different type of dyes decolourization, with the addition of veratryl alcohol [210].

In the case of VP from *B. fumosa* the addition of *p*-aminobenzoic acid, Tween-80, anthracene and fluorene caused reversible inhibition of VP, and the presence of NaN_3 caused irreversible inhibition of enzyme [216]. Activity of VP can also be inhibited by Ca^{2+} depletion by alkaline or thermal treatment. Ca^{2+} is VP's structural element, important for stabilize of proximal and distal histidines in the enzyme molecule what is very important during the initial step of VP's catalytic cycle [227].

Mutual inhibition between oxidation of the substrates Reactive Black 5 VP from *P. eryngii* (an azo dye that is not oxidized by Mn^{3+}) and Mn^{2+} was found. However, the azo dye appeared as a stronger inhibitor because the inhibitor/substrate molar ratio, to cause 40% inhibition, was 12% in the case of Mn^{2+} inhibition of dye oxidation, and only 0.13 in the case of Reactive Black 5 inhibition of Mn^{2+} oxidation. The type of inhibition caused by Reactive Black 5 was investigated using different dye concentrations, and although a decrease of the velocity of Mn^{2+} oxidation was found, the K_M value was maintained. This suggests two different substrate interaction sites with different affinities that are not affected by the presence of the alternative substrate acting as inhibitor. The inhibitory effect would be caused by the reduction of the oxidized enzyme forms (Compounds I and II) during oxidation of the substrate acting as inhibitor. Finally, it is possible to mention that the stronger inhibitory effect caused by Reactive Black 5 agrees with the higher peroxidase affinity for this substrate, compared with the affinity for Mn^{2+} [219]. Reactive Black 5 acts as a very efficient noncompetitive inhibitor of Mn^{2+} oxidation by the *P. eryngii* VP. This result agrees with the noncompetitive inhibition by Mn^{2+} of the oxidation of dyes and veratryl alcohol by two similar peroxidases recently isolated from *Pleurotus* and *Bjerkandera* species [219].

5. The Degradative Properties of Ligninolytic Fungi As Result of Unique Properties of Peroxidases

In summary, according to the presented data the following remarks should be made. The degradative spectrum of white rot and litter-decomposing fungi is the same and include lignin, humic substances, dyes, dioxins, chlorinated compounds, pharmaceutical compounds, endocrine-disrupting compounds, trinitrotoluene, and polycyclic aromatic hydrocarbons (Table 1). The set of substrates which can be degraded by LiP-MnP-Lac producing fungi and MnP-Lac producing fungi is the same. Now little information was found on degradative activity of the fungi belonging to III and IV groups. The most active decomposers are the members of first (LiP-MnP-Lac) and second (MnP-Lac) groups. The mineralization activity was found among of the members of I and II group too.

Comparing the data presented in Table 1-4 it should noted that the compounds, which are the substrate of the fungi, also are the substrates of the ligninolytic peroxidases. It was found that ligninolytic peroxidases can be involved both to first attack of lignin [26, 31, 38, 53, 56, 114, 116] and xenobiotics [4, 5, 7, 8, 117, 118] and to steps of utilization of the products of degradation [11, 12]. In spite of many differences of properties all studied peroxidases can catalyze the depolimerization of lignin [41, 195, 224], oxidative dehalogenation reaction [5, 214] and the polymerization of monomeric compounds [41]. As we can see from Table 2-4 the main range of substrates oxidized by all ligninolytic peroxidases is the same.

In addition some unique properties of different peroxidases should be mentioned. For example, LiP has the unique ability to catalyze oxidative cleavage of carbon-carbon bonds and ether bonds (C-O-C) in high-redox potential non-phenolic aromatic substrates [123, 124]. The key functional possibility of LiP is that this ligninolytic peroxidase can oxidize aromatic rings that are only moderately activated by electron-donating substituents [129]. Reductive dehalogenation of some aliphatic halocarbons by LiP was found too [153]. It is worthy to note that LiP as strong oxidant (in comparison to other ligninolytic enzymes) does not require the participation of mediators in the oxidation of non-phenolic substrates but is not able to oxidize phenolic compounds efficiently because of inactivation in the absence of veratryl alcohol. The cation radical of veratryl alcohol acts as redox mediator and a similar role has been suggested for oxidation of some non-phenolic aromatic substrates [25]. Other a natural compounds produced *denovo* by several white rot fungi, such as 2-chloro-1,4-dimethoxybenzene, may be mediators of different substrates oxidation superior to veratryl alcohol [137].

Despite the MnP/Mn^{2+} couple is not able to oxidize non-phenolic lignin model dimers as LiP does, partial depolymerisation of synthetic and plant lignin has been reported probably by endwise attack progressing from phenolic units. Other studies expanded the role of MnP in lignin biodegradation via thiol and lipid-derived free radicals that are able to oxidize a variety of non-phenolic aromatic compounds [25]. Extracellular membranes are frequently attached to the hyphae of white rot fungi, and peroxidases are found on these structures. The chemical composition of these extracellular membranes indicated that they contain lipids. An analysis of the extractable lipids in *P. chrysosporium* mycelium showed that they contain unsaturated fatty acids. Lipid peroxidation is frequently initiated by transition metalions that react with endogenous lipid hydroperoxides, either oxidizing them to peroxy radicals or reducing them to alkoxyradicals. In the case of MnP from *P. chrysosporium*, possible initiators include Mn^{2+}, Mn^3+, and the peroxidase heme. The oxygen-centered radicals produced during lipid peroxidation are known to trigger xenobiotic cooxidations. For example, phenanthrene is most likely to be oxidized by the latter route [191]. The direct involvement of MnP in the mineralization of PAHs [14, 187, 197], pentachlorophenol, catechol, tyrosine, tryptophane, and 2-amino-4,6-dinitrotoluene [194, 202] was found too. The MnP-catalyzed oxidation of these compounds leaded to "enzymatic combustion" (the formation of $^{14}CO_2$) [197]. In addition MnP from *P. radiata* and *N. frowardii* was found to decompose malonate oxidatively in the absence of H_2O_2. The enzymatic oxidation resulted in a substantial decrease in malonate concentration and the formation of CO_2, oxalate, glyoxylate and formate [204].

As ligninolytic peroxidase VP is capable of degrading substrates specific for MnP and LiP. The different catalytic sites of this enzyme are responsible for the oxidation of low- and high-redox-potential compounds. Additionally VP can oxidize a wide range of substrates because of its high redox potential as well [1]. It is worth noting that VP's structure is more relating to the structure of LiP (58-60%) than to the structure of MnP (55%) [210]. Unlike general peroxidases, VP from *P. ostreatus* has a unique property of direct oxidization of high-molecular-weight compounds, such as Poly R-478 and RNase A [225]. During VP-mediated oxidation of some substrates, both with and without Mn^{2+}, biphasic kinetics with apparent saturation in both micromolar and millimolar ranges were obtained (Table 4). Similar biphasic kinetic have also been reported for the oxidation of selected substrates by LiP [73].

In conclusion the catalytic properties required for oxidation of high-redox potential and bulky molecules, are provided by several unique characteristics in the molecular architecture of ligninolytic peroxidases such as the presence of oxidation sites on the protein surface to collect electrons from substrates and deliver them to heme using different pathways. In addition the catalytic possibilities of LiP can be extended to phenolic compounds by using of natural mediator and catalytic possibilities of MnP can be extended to non-phenolic compounds by thiol and lipid-peroxidation derived free radicals.

6. Biotechnological Uses of Peroxidases As Catalysts and During Mycoremediation Processes

Due to the high redox-potential required for lignin oxidation, ligninolytic peroxidases have been the object of high interest as industrial biocatalysts [1]. In recent years, the application of ligninolytic peroxidases to the degradation of lignin and xenobiotics such as PAHs [13-19], dyes [3, 4] and a wide range of pesticides [5, 6] has been intensively studied. For example, convertion in vitro milled wood by MnP from *P. radiata* resulted in the polymerization of lower-molecular-mass, soluble wood components and in the partial depolimerization of the insoluble bulk of pine wood [228]. A pretreatment of wheat straw by LiPs from different white rot fungi increased the following saccharification yield of this substrate [229].

Among the different basidiomycete peroxidases, VP presents particular interest due to its catalytic versatility including the degradation of compounds that other peroxidases are not able to oxidize directly. The ability to degrade directly a variety of recalcitrant compounds represents a VP characteristic of the highest biotechnological interest. The use of aromatic mediators acting via their cation radicals is required for LiP oxidation of simple phenols, monomethoxylated aromatics, polymeric dyes, PAHs, and, most important, polymeric lignin. By contrast, VP does not require mediators to oxidize many of the above compounds, including phenolic and non-phenolic aromatic pollutants, pesticides, high redox-potential and polymeric dyes, PAHs, and lignin. This VP ability represents an important advantage with respect to LiP, since the cost of VA as a mediator can be saved. Moreover, for more efficient oxidation of some recalcitrant compounds VP can also be used in combination with redox mediators including natural phenolic compounds [1]. VP versatility permits its application in Mn^{3+}-mediated or Mn-independent reactions on both low and high redox-potential aromatic substrates and dyes, among others. It has been shown that VP can be used to reoxidize Mn-containing polyoxometalates, which are efficient oxidizers in paper pulp delignification. Another aspect of VP catalysis, which has been reported recently and merits investigation in more detail for some applications, is its high activity in the presence of organic solvents [1, 206].

It is known that free enzyme is too fragile to resist harsh operational conditions, such as extremely low or high pH values and high temperatures. As a result, enzyme cost is a major hurdle and thus, enzyme immobilization is desirable for solving these problems and increasing the efficiency of the enzyme [230]. Only inexpensive enzymes could be used in the soluble form, since treatment processes would require large quantities of enzyme that would be lost in the effluent [64]. Insoluble enzyme preparations are useful catalysts for industrial biotransformations. Being more easily separated and recycled than their soluble counterparts, they are conductive to continuous process development and are therefore often the preferred choice of biocatalysts for manufacturing scale process operation [230]. Immobilized proteins have been used in a wide variety of applications including production of special chemicals, stereoselective synthesis, as biocatalysts in the degradation of toxic organic compounds etc. Oxidative enzymes from ligninolytic fungi are diverse biocatalysts with tremendous industrial potential for transformations and mineralizations of recalcitrant xenobiotics including phenols, amines, organic dyes, methoxybenzenes, PAHs, and complex mixtures of chlorinated lignin. In order to take full advantage of ligninolytic enzymes as biocatalysts for organopollutants degradation and minimize the disadvantages of employing the free soluble protein, these enzymes must be immobilized [64].

Now the use of immobilized ligninolytic peroxidases is not well documented. For example, immobilized LiP from *Chrisonilia sitophila* decolorizes phenolic kraft effluent from pulp mills [231]. Covalently immobilized MnP from *Lentinula edodes* was capable of generating Mn^{3+} from Mn^{2+} and H_2O_2. Mn^{3+}, properly chelated, can be used as a nonspecific oxidant of organopollutants [64]. Immobilization enhanced activity LiP from *T. versicolor* IBL-04, stability over broad pH range (3-8) and thermotolerance in 20-80°C temperature range. Immobilization also enhanced the substrate affinity and catalytic efficiency of LiP as evidenced by lower K_M and high V_{max} values in veratryl alcohol oxidation. Immobilized LiP had significantly higher tolerance against inactivating agents (cysteine, EDTA, Ag^+) as compared to free enzyme [232]. Native and PEG-modified MnP from *B. adusta* UAMH 8258 showed similar catalytic properties in the oxidation of Mn^{2+} and other substrates but modified enzyme showed greater resistance to denaturation by hydrogen peroxide and stability to organic solvents, to higher temperatures and lower pH [220].

The environment is continuously polluted by a large array of hazardous chemicals with different structures and different toxicity levels that are released from several anthropogenic sources. The use of the living organisms (bioremediation) to remediate polluted environments is a promising and growing area of environmental biotechnology. The use of extracellular enzymes has been proposed as an innovative remediation technique [233]. Extracellular non-specific enzyme system of the ligninolytic fungi permits them to metabolize the wide range of xenobiotics which are not-available and hard-available to soil microflora. Partial oxidation of recalcitrant pollutants such as PAHs by extracellular enzymes give rise to products of increased polarity and water solubility and thus with a higher biodegradability to soil microorganisms [233].

The ligninolytic enzymes are glycoproteins [20] and are embedded within extracellular polysaccharide sheets around the hyphae of wood-rotting basidiomycetes, which have been proposed to be involved in recognition mechanisms in fungal cell-wood surface interactions [234]. These two properties can stabilized the enzymes in soil. Enzyme-containing sheets around fungal cells may also effect the spatial partitioning and hence the bioavailability of pollutants for degrading fungi.

Of course, the treatment of contaminated soil by pure enzymes is nonsense but some optimal conditions for its production and operation can be made. As demonstrated in *P.*

chrysosporium [235] ligninolytic peroxidases are usually expressed under nutrient-deficient conditions that are prevalently encountered in most contaminated soils. The introduction of the fungi on the natural lignin-containing substrates (sawdust, straw, etc.) can serve carbon soure for growth of the fungi and the products of its degradation can induced the ligninolytic peroxidases. The addition of surfactants, like Tween-80, can also affect better solvation of the hydrophobic substrates and provide the coupling of oxidation and lipid-peroxidation reactions of MnP.

In a polluted site, pollutant cocktails, normally composed of combinations of many organic (and inorganic) contamitants, rather than a single polluting substance are present. As a consequence possible negative, or positive, synergistic effects on the enzyme efficiency may occur [233]. In the soil, enzymes are present in complex, three-dimensional assemblages of mineral and organic particles that will restrict their mobility and affect their activity. Enzyme molecules can be adsorbed, immobilized, or entrapped in such matrices giving rise to the so-called "naturally immobilized enzymes". Inactivation or degradation of enzymatic molecules may also take place. As a consequence, change in their kinetic, stability and mobility will occur [233].

Finally, white rot and litter-decomposing fungi can actively degrade xenobiotics as in a model soil systems as in the conditions of the real soil, for example in soils with high salinity and in soils containing concomitant contamination such as heavy metals. The decrease of the toxicity of the soil during mycoremediation makes this method perspective for the use.

Conclusion

The presented review is an attempt to summarize the data on degradative activity of different groups of ligninolytic fungi (white rot and litter-decomposing) according to properties of producing peroxidases. We propose the separation of ligninolytic fungi to six group according to set of producing enzymes: I – LiP-MnP-Lac, II – MnP-Lac, III – VP-MnP-Lac, IV – LiP-MnP-VP, V – LiP-Lac, and VI – the fungi with insufficient information. The production of ligninolytic peroxidases vary and dependent from fungi species and cultivation conditions. This dependence is correct to fungi belonging to different group.

The induction of the production of LiP and MnP by Mn^{2+} ions is common property of white rot and litter-decomposing fungi with the exception of *Pleurotus* species. The fungal cultures can produce some metabolites such as veratryl alcohol [49, 109, 110] and water-soluble products of lignin degradation [111, 112] which are a good inductors of ligninolytic peroxidases production.

LiP is not commonly produced by ligninolytic fungi. Litter-decomposing fungi, which did not produce LiP, are able to degrade and effective mineralize recalcitrant biopolymers. The lack of LiP suggests that the fungi may use alternative mechanisms of degradation. Evidence for the critical role of MnP in lignin and xenobiotics degradation is accumulating. The direct involvement of MnP in the mineralization of PAHs [14, 187, 197], pentachlorophenol, catechol, tyrosine, tryptophane, and 2-amino-4,6-dinitrotoluene [194, 202] was found too. The catalytic possibilities of both enzymes can be extended (LiP by natural mediator and MnP by thiol and lipid-peroxidation). Perhaps at the absence of LiP the function of first attack of hard moleculs can realize MnP. In this case the different mechanisms such as coupling of oxidation and lipid-peroxidation can occur.

It can be proposed that MnP according to its catalytic properties is the most important peroxidase among ligninolytic peroxidases. VP, which combines the properties of LiP and MnP, can be more "evolutionary advanced" enzyme, but now VP was found in members of two genus (*Bjerkandera* and *Pleurotus*). Some confusion is in consideration of thise enzyme. Some authors [98, 220] described the catalytic properties of MnPs from *P. ostreatus* and *B. adusta*; in fact these peroxidases are VP. MnP produced by *B. adusta* UAMH 8258 has Mn^{2+}-dependent and independent activities [220]. It is not clear if these fungi have two enzymes (MnP and VP) or only VP.

In conclusion the unique catalytic properties of LiP, MnP, and VP confirm its key role in degradation of lignin and xenobiotics by ligninolytic fungi and make the ligninolytic peroxidases perspective to biotechnological uses.

References

[1] Ruiz-Duenas, F.J., Pogni, R., Morales, M., Giansanti, S., Mate, M.J., Romero, A., Martinez, M.J., Basosi, R., Martinez, A.T. (2009) Protein radicals in fungal versatile peroxidase: Catalytic tryptophan radical in both Compound I and Compound II and studies on W164Y, W164H, and W164S variants. *J. Biol. Chem., 284,* 7986-7994.

[2] Snajdr, J., Steffen, K.T., Hofrichter, M., Baldrian, P. (2010) Transformation of ^{14}C-labelled lignin and humic substances in forest soil by the saprobic basidiomycetes *Gymnoporus erythropus* and *Hypholoma fasciculare. Soil Biol. Biochem., 42,* 1541-1548.

[3] Wesenberg, D., Kyriakides, I., Agathos, S.N. (2003) White rot fungi and their enzymes for the treatment of industrial dye effluents. *Biotechnol. Adv., 22,* 161-187.

[4] Harazono, K., Watanabe, Y., Nakamura, K. (2003) Decolorization of azo dye by the white rot basidiomycetes *Phanerochaete sordida* and its manganese peroxidase. *J. Bioscience Bioeng., 95,* 455-459.

[5] Valli, K., Gold, M. (1991) Degradation of 2,4-dichlorophenol by lignin degrading fungus *Phanerochaete chrysosporium. J. Bacteriol., 173,* 345-352.

[6] Suhara, H., Adachi, A., Kamei, I., Maekawa, N. (2011) Degradation of chlorinated pesticide DDT by litter-decomposing basidiomycetes. *Biodegradation, 22,* 1075-1086.

[7] Golan-Rozen, N., Chefetz, B., Ben-Ari, J., Geva, J., Hadar, Y. (2011) Transformation of the recalcitrant pharmaceutical compound carbamazepine by *Pleurotus ostreatus*: Role of cytochrome P450 monooxygenase and manganese peroxidase. *Environ. Sci. Technol., 45,* 6800-6805.

[8] Shin, E.-H., Choi, H.T., Song, H.-G. (2007) Biodegradation of endocrine-disrupting bisphenol A by white rot fungus *Irpex lacteus. J. Microbiol. Biotechnol., 17,* 1147-1151.

[9] Soon-Seok, H., Choi, H.T., Song, H.-G. (2008) Biodegradation of endocrine-disrupting phthalates by *Pleurotus ostreatus. J. Microbiol. Biotechnol., 18,* 767-772.

[10] Kabiersch, G., Rajasarkka, J., Ullrich, R., Tuomela, M., Hofrichter, M., Virta, M., Hatakka, A., Steffen, K. (2011) Fate of bisphenol A during treatment with the litter-

decomposing fungi *Stropharia rugosoannulata* and *Stropharia coronilla*. *Chemosphere, 83,* 226-232.

[11] Aken, B., Skubisz, K., Naveau, H., Agathos, S.N. (1997) Biodegradation of 2,4,6-trinitrotoluene (TNT) by the white rot basidiomycetes *Phlebia radiata*. *Biotechnol. Lett., 19,* 813-817.

[12] Kim, H.-Y., Song, H.-G. (2000) Transformation of 2,4,6-trinitrotoluene by white rot fungus *Irpex lacteus*. *Biotechnol. Lett., 22,* 969-975.

[13] Schutzendubel, A., Majcherczyk, A., Johannes, C., Huttermann, A. (1999) Degradation of fluorene, anthracene, phenanthrene, fluoranthene and pyrene lacks connection to the production of extracellular enzymes by *Pleurotus ostreatus* and *Bjerkandera adusta*. *Int. Biodeter. Biodegr., 43,* 93-100.

[14] Steffen, K.T., Hatakka, A., Hofrichter, M. (2003) Degradation of benzo[a]pyrene by the litter-decomposing basidiomycete *Stropharia coronilla*: Role of manganese peroxidase. *Appl. Environ. Microbiol., 69,* 3957-3964.

[15] Han, M.-J., Choi, H.-T., Song, H.-G. (2004) Degradation of phenanthrene by *Trametes versicolor* and its laccase. *J. Microbiol., 42,* 94-98.

[16] Cajthaml, T., Erbanova, P., Sasek, V., Moeder, M. (2006) Breakdown products on metabolic pathway of degradation of benz[a]anthracene by a ligninolytic fungus. *Chemosphere, 64,* 560-564.

[17] Chupungars, K., Rerngsamran, P., Thaniyavarn, S. (2009) Polycyclic aromatic hydrocarbons degradation by *Agrocybe* sp. CU-43 and its fluorene transformation. *Int. Biodeter. Biodegr., 63,* 93-99.

[18] Lee, H., Choi, Y.-S., Kim, M.-J., Huh, N.-Y., Kim, G.-H., Lim, Y.W., Kang, S.-M., Cho, S.-T., Kim, J.-J. (2010) Degradation ability of oligocyclic aromates by *Phanerochaete sordida* selected via screening of white rot fungi. *Folia Microbiol., 55,* 447-453.

[19] Pozdnyakova, N.N., Nikiforova, S.V., Makarov, O.E., Chernyshova, M.P., Pankin, K.E., Turkovskaya, O.V. (2010) Influence of cultivation conditions on pyrene degradation by the fungus *Pleurotus ostreatus* D1. *World J. Microbiol. Biotechnol., 26,* 205-211.

[20] Wong, D.W.S. (2009) Structure and action mechanism of ligninolytic enzymes. *Appl. Biochem. Biotechnol., 157,* 174-209.

[21] Ertan, H., Siddiqui, K., Muenchhoff, J., Charlton, T., Cavicchioli, R. (2012) Kinetic and thermodynamic characterization of the functional properties of a hybrid versatile peroxidase using isothermal titration calorimetry: insight into manganese peroxidase activation and lignin peroxidase inhibition. *Biochimie, 94,* 1221-1231.

[22] Glenn, J.K. and Gold, M.H. (1983) Decolorization of several polymeric dyes by the lignin-degrading basidiomycete *Phanerochaete chrysosporium*. *Appl. Environ. Microbiol., 45,* 1741-1747.

[23] Gold M.H., Kuwahara M., Chiu A.A., Glenn J.K. (1984) Purification and characterization of an extracellular H_2O_2-requiring diarilpropane oxygenase from the white rot basidiomycete *Phanerochaete chrysosporium*. *Arch. Biochem. Biophys., 234,* 353-362.

[24] Kuwahara, M., Glenn, J.K., Morgan, M.A., Gold, M.H. (1984) Separation and characterization of two extracellular H_2O_2-dependent oxidases from ligninolityc cultures of *Phanerochaete chrysosporium*. *FEBS Lett., 169,* 247-250.

[25] Martinez, A. (2002) Molecular biology and structure-function of lignin-degrading heme peroxidases. *Enzyme Microbial. Technol., 30,* 425-444.

[26] Hatakka, A. (1994) Lignin modifying enzymes from selected white rot fungi: production and role in lignin degradation. *FEMS Microbiol. Rev., 13,* 125-135.

[27] Jarosz-Wilkolazka, A., Luterek, J., Olszewska, A. (2008) Catalytic activity of versatile peroxidase from *Bjerkandera fumosa* at different pH. *Biocatal Biotransform, 26,* 280-287.

[28] Zorn, H., Langhoff, S., Scheibner, M., Nimtz, M., Berger, R. (2003) A peroxidase from *Lepista irina* cleaves *β,β*-carotene to flavor compounds. *Biol. Chem., 384,* 1049-1056.

[29] Lisov, A., Leontievsky, A., Golovleva, L. (2003) Hybrid Mn-peroxidase from the ligninolytic fungus *Panus tigrinus* 8/18. Isolation, substrate specificity, and catalytic cycle. *Biochemistry (Moscow), 68,* 1027-1035.

[30] Wesenberg, D., Buchon, F., Agathos, S.N. (2002) Degradation of dye-containing textile effluent by the agaric white rot fungus *Clitocybula dusenii. Biotechnol. Lett., 24,* 989-993.

[31] Grinhut, T., Hadar, Y., Chen, Y. (2007) Degradation and transformation of humic substances by saprotrophic fungi: processes and mechanisms. *Fungal Biol. Rev., 21,* 179-189.

[32] Kawai, S., Umezawa, T., Higuchi, T. (1985) Metabolism of a non-phenolic *β*-O-4 lignin substructure model compound by *Coriolus versicolor. Agric. Biol. Chem., 49,* 2325-2330.

[33] Jönsson, L., Karlsson, O., Lundquist, K., Nyman, P.O. (1989) *Trametes versicolor* ligninase: isozyme sequence homology and substrate specificity. *FEBS Lett., 247,* 143-146.

[34] Katayama, Y., Nishida, T., Morohoshi, N., Kuroda, K. (1989) The methabolism of biphenyl structures in lignin by the wood-rotting fungus *Coriolus versicolor. FEMS Microbiol. Lett., 61,* 307-313.

[35] Silva, C.M.M.S., de Melo, I.S., de Oliveira, P.R. (2005) Ligninolytic enzyme production by *Ganoderma* spp. *Enzyme Microbial. Technol., 37,* 324-329.

[36] Song, H.-G. (1997) Biodegradation of aromatic hydrocarbons by several white rot fungi. *J. Microbiol., 35,* 66-71.

[37] Rothschild, N., Novotny, C., Šašek, V., Dosoretz, C.G. (2002) Ligninolytic enzymes of the fungus *Irpex lacteus* (*Polyporus tulipiferae*): isolation and characterization of lignin peroxidase. *Enzyme Microbial Technol., 31,* 627-633.

[38] Hofrichter, M., Vares, T., Kalsi, M., Galkin, S., Scheibner, K., Fritsche, W., Hatakka, A. (1999) Production of manganese peroxidase and organic acids and mineralization of ^{14}C-labelled lignin (^{14}C-DHP) during solid-state fermentation of wheat straw with the white rot fungus *Nematoloma frowardii. Appl. Environ. Microbiol., 65,* 1864-1870.

[39] Hilden, K.S., Bortfeldt, R., Hofrichter, M., Hatakka, A., Lundell, T.K. (2008) Molecular characterization of the basidiomycetes isolate *Nematoloma frowardii* b19 and its manganese peroxidase places the fungus in the corticioid genus *Phlebia. Microbiology, 154,* 2371-2379.

[40] Dhawale, S.W., Dhawale, S.S., Dean-Ross, D. (1992) Degradation of phenanthrene by *Phanerochaete chrysosporium* occurs under ligninolytic as well as nonligninolytic conditions. *Appl. Environ. Microbiol., 58,* 3000-3006.

[41] Ruttimann-Johnson, C., Lamar, R.T. (1996) Polymerization of pentachlorophenol and ferulic acid by fungal extracellular lignin-degrading enzymes. *Appl. Environ. Microbiol., 62,* 3890-3893.

[42] Vyas, B.R.M., Bakowski, S., Sasek, V., Matucha, M. (1994) Degradation of anthracene by selected white rot fungi. *FEMS Microbiol Ecol., 14,* 65-70.

[43] Chung, N., Aust, S. (1995) Inactivation of lignin peroxidase during phenol oxidation. *Arch. Biochem. Biophys., 316,* 851-856.

[44] Bogan, B.W., Lamar, R.T. (1996) Polycyclic aromatic hydrocarbon degrading capabilities of *Phanerochaete laevis* HHB-1625 and its extracellular ligninolytic enzymes. *Appl. Environ. Microbiol., 62,* 1597-1603.

[45] Deguchi, T., Kitaoka, Y., Kakezawa, M., Nishida, T. (1998) Purification and characterization of a nylon-degrading enzyme. *Appl. Environ. Microbiol., 64,* 1366-1371.

[46] Niku-Paavola, M.L., Karhunen, E., Salola, P., Raunio, V. (1988) Ligninolytic enzymes of the white rot fungus *Phlebia radiata. Biochem. J., 254,* 877-884.

[47] Vares, T., Niemenmaa, O., Hatakka, A. (1994) Secretion of ligninolytic enzymes and mineralization of ^{14}C-ring-labelled synthetic lignin by three *Phlebia tremellosa* strains. *Appl. Environ. Microbiol., 60,* 569-575.

[48] Nyanhongo, G.S., Gomes, J., Gűbitz, G.M., Zvauya, R., Read, J., Steiner W. (2002) Decolorization of textile dyes by laccases from a newly isolated strain of *Trametes modesta. Wat. Res., 36,* 1449-1453.

[49] Rogalski, J., Lundell, T., Leonowicz, A., Hatakka, A. (1991) Production of laccase, lignin peroxidase and manganese-dependent peroxidase by various strains of *Trametes versicolor* depending on culture conditions. *Acta Microbiol. Pol., 40,* 221-234.

[50] Perry, C., Matcham, S., Wood, D., Thurston, C. (1993) The structure of laccase protein and its synthesis by the commercial mushroom *Agaricus bisporus. J. Gen. Microbiol., 139,* 171-178.

[51] ten Have, R., Wijngaard, H., Aris-Kronenburg, N., Straatsma, G., Schaap, P.J. (2003) Lignin degradation by *Agaricus bisporus* accounts for a 30 increase in bioavailable holocellulose during cultivation on compost. *J. Agric. Food Chem., 51,* 2242-2245.

[52] Lankinen, P., Hildén, K., Aro, N., Salkinoja-Salonen, M., Hatakka, A. (2005) Manganese peroxidase of *Agaricus bisporus*: grain bran-promoted production and gene characterization. *Appl. Microbiol. Biotechnol., 66,* 401–407.

[53] Daina, S., Orlandi, M., Bestetti, G., Wiik, C., Elegir, G. (2002) Degradation of *β*-5 lignin model dimers by *Ceriporiopsis subvermispora. Enzyme Microbial. Technol., 30,* 499-505.

[54] Rogalski, J., Dawidowicz, A., Jozwik, E., Leonowicz, A. (1999) Immobilization of laccase from *Cerrena unicolor* on controlled porosity glass. *J. Mol. Catalysis B: Enzymatic, 6,* 29-39.

[55] Koroleva, O.V., Gavrilova, V.P., Stepanova, E.V., Lebedeva, V.I., Sverdlova, N.I., Landesman, E.O., Yavmetdinov, I.S., Yaropolov, A.I. (2002) Production of lignin modifying enzymes by co-cultivated white rot fungi *Cerrena maxima* and *Coriolus hirsutus* and characterization of laccase from *Cerrena maxima. Enzyme Microbial Technol., 30,* 573-580.

[56] Steffen, K.T., Hofrichter, M., Hatakka, A. (2002) Degradation of humic acids by the litter-decomposing basidiomycetes *Collibia dryophila. Appl. Environ. Microbiol., 68,* 3442-3448.
[57] Scheel, T., Hofer, M., Ludwig, S., Holker, U. (2000) Differential expression of manganese peroxidase and laccase in white rot fungi in the presence of manganese or aromatic compounds. *Appl. Microbiol. Biotechnol., 54,* 686-691.
[58] Gill, P.K., Arora, D.S., Chander, M. (2002) Biodecolourization of azo and triphenylmethane dyes by *Dichomitus squalens* and *Phlebia* spp. *J. Industrial Microbiol. Biotechnol., 28,* 201-203.
[59] Papinutti, L., Martinez, M.J. (2006) Production and characterization of laccase and manganese peroxidase from the ligninolytic fungus *Fomes sclerodermeus. J. Chem. Technol. Biotechnol., 81,* 1064-1070.
[60] D'Souza, T.M., Merritt, C.S., Reddy, C.A. (1999) Lignin-modifying enzymes of the white rot basidiomycete *Ganoderma lucidum. Appl. Environ. Microbiol., 65,* 5307-5313.
[61] Murugesan, K., Kim, Y.-M., Jeon, J.-R., Chang, Y.-S. (2009) Effect of metal ions on reactive dye decolorization by laccase from *Ganoderma lucidum. J. Hazard. Materials, 168,* 523-529.
[62] Ting, W.T.E., Yuan, S.Y., Wu, S.D., Chang, B.V. (2011) Biodegradation of phenanthrene and pyrene by *Ganoderma lucidum. Int. Biodeter. Biodegr., 65,* 238-242.
[63] Gramss, G., Ziegenhagen, C., Sorge, S. (1999) Degradation of soil humic extract by wood- and soil-associated fungi, bacteria, and commercial enzymes. *Microbial. Ecol., 37,* 140-151.
[64] Grabski, A., Coleman, P., Drtina, G., Burgess, R. (1995) Immobilization of manganese peroxidase from *Lentinula edodes* on azlactone-functional polymers and generation of Mn^{3+} by the enzyme-polymer complex. *Appl. Biochem. Biotechnol., 55,* 55-73.
[65] Heinzkill, M., Bech, L., Halkier, T., Schneider, P., Anke, T. (1998) Characterization of laccases and peroxidases from wood-rotting fungi (family *Coprinaceae*). *Appl. Environ. Microbiol., 64,* 1601-1606.
[66] Perez, J., de la Rubia, T., ben Hamman, O., Martinez, M.J. (1998) *Phanerochaete flavido-alba* laccase induction and modification of manganese peroxidase isoenzyme pattern in decolorized olive oil mill wastewaters. *Appl. Environ. Microbiol., 64,* 2726-2729.
[67] Hofrichter, M. (2002) Review: Lignin conversion by manganese peroxidase (MnP). *Enzyme Microb. Tech., 30,* 454-466.
[68] Garzillo, A.M., Colao, M.C., Buonocore, V., Oliva, R., Falcigno, L., Savino, M., Santoro, A.M., Zappala, R., Bonomo, R.P., Bianco, C., Giardina, P., Palmieri, G., Sannia, G.J. (2001) Structural and kinetic characterization of native laccase from *Pleurotus ostreatus, Rigidoporus lignosus* and *Trametes trogii. Protein Chem., 20,* 191-201.
[69] Steffen, K.T., Hatakka, A., Hofrichter, M. (2002) Removal and mineralization of polycyclic aromatic hydrocarbons by litter-decomposing basidiomycetous fungi. *Appl. Microbiol. Biotechnol., 60,* 212-217.
[70] Kapich, A.N., Steffen, K.T., Hofrichter, M., Hatakka, A. (2005) Involvement of lipid peroxidation in the degradation of a non-phenolic lignin model compounds by

manganese peroxidase of the litter-decomposing fungus *Stropharia coronilla. Biochem. Biophys. Res. Commun., 330,* 371-377.

[71] Maltseva, O.V., Niku-Paavola, M.-L., Leontievsky, A.A., Myasoedova, N.M., Golovleva, L.A. (1991) Ligninolytical enzymes of the white rot fungus *Panus tigrinus. Biotechnol. Appl. Biochem., 13,* 291-302.

[72] Munoz, C., Guillen, F., Martınez, A.T., Martınez, M.J. (1997) Induction and characterization of laccase in the ligninolytic fungus *Pleurotus eryngii. Curr. Microbiol., 34,* 1-5.

[73] Ruiz-Duenas, F.J., Camarero, S., Perez-Boada, M., Martinez, M.J., Martinez, A.T. (2001) A new versatile peroxidase from *Pleurotus. Biochem. Soc. Transact., 29,* 116-122.

[74] Velazquez-Cedeno, M.A., Mata, G., Savoie, J.-M. (2002) Waste-reducing cultivation of *Pleurotus ostreatus* and *Pleurotus pulmonarius* on coffee pulp: changes in the production of some lignocellulolytic enzymes. *World J. Microbiol. Biotechnol., 18,* 201-207.

[75] Muheim, A., Leisola, M.S.A., Schoemaker, E. (1990) Aryl-alcohol oxidase and lignin peroxidase from the white rot fungus *Bjerkandera adusta. J. Biotechnol., 13,* 159-167.

[76] ten Have, R., Hartmans, S., Teunissen, P.J., Field, J.A. (1998) Purification and characterization of two lignin peroxidase isozymes produced by *Bjerkandera* sp. strain BOS55. *FEBS Lett., 422,* 391-400.

[77] Mielgo, I., Lopez, C., Moreira, M.T., Feijoo, G., Lema, J.M. (2003) Oxidative degradation of azo dyes by manganese peroxidase under optimized conditions. *Biotechnol. Prog., 19,* 325-331.

[78] Ullrich, R., Huong, L.M., Dung, N.L., Hofrichter, M. (2005) Laccase from the medicinal mushroom *Agaricus blazei*: production, purification and characterization. *Appl. Microbiol. Biotechnol., 67,* 357-363.

[79] Rehman, A.V., Thurston, C.F. (1992) Purification of laccase from *Armillaria mellea. J.Gen. Microbiol., 138,* 1251-1257.

[80] Li, K., Xu, F., Eriksson, K.-E. (1999) Comparison of fungal laccases and redox mediators in oxidation of a nonphenolic lignin model compound. *Appl. Environ. Microbiol., 65,* 2654-2660.

[81] Fukushima, Y., Kirk, T.K. (1995) Laccase component of the *Ceriporiopsis subvermispora* lignin-degrading system. *Appl. Environ. Microbiol., 61,* 872-876.

[82] Schneider, P., Caspersen, M., Mondorf, K., Halkier, T., Skov, L.K., Ostergaard, P.R., Brown, K.M., Brown, S.H., Xu, F. (1999) Characterization of *Coprinus cinereus* laccase. *Enzyme Microbial. Technol., 25,* 502-508.

[83] Saparrat, M.C.N., Guillén, F., Arambarri, A.M., Martínez, A.T., Martínez, M.J. (2002) Induction, isolation, and characterization of two laccases from the white rot basidiomycete *Coriolopsis rigida. Appl. Environ. Microbiol., 68,* 1534-1540.

[84] Xing, Z.T., Cheng, J.H., Tan, Q., Pan, Y.J. (2006) Effect of nutritional parameters on laccase production by the culinary and medicinal mushroom *Grifola frondosa. World J. Microbiol. Biotechnol., 22,* 799-806.

[85] Tagger, S., Perissol, C., Gill, G., Vogt, G., le Petit, J. (1998) Phenoloxidases of the white rot fungus *Marasmius quercophilus* isolated from an evergreen oak litter (*Quercus ilex* L.). *Enzyme Microbial Technol., 23,* 372-379.

[86] Linke, D., Bouws, H., Peters, T., Nimtz, M., Berger, R.G., Zorn, H. (2005) Laccases of *Pleurotus sapidus*: Characterization and cloning. *J. Agric. Food Chem., 53,* 9498-9505.
[87] Schliephake, K., Mainwaring, D.E., Lonergan, G.T., Jonesa, I.K., Bakera, A., Warren, L. (2000) Transformation and degradation of the disazo dye Chicago Sky Blue by a purified laccase from *Pycnoporus cinnabarinus*. *Enzyme Microbial Technol., 27,* 100-107.
[88] Bonomo, R.P., Cennamo, G., Purrello, R., Santoro, A.M., Zapala, R. (2001) Comparison of three fungal laccases from *Rigidoporus lignosus* and *Pleurotus ostreatus*: correlation between conformation chenges and catalytic activity. *J. Inorg. Biochem., 83,* 67-75.
[89] Galhaup, C., Wagner, H., Hinterstoisser, B., Haltricha, D. (2002) Increased production of laccase by the wood-degrading basidiomycete *Trametes pubescens*. *Enzyme Microbial Technol., 30,* 529-536.
[90] Chen, S., Ge, W., Buswell, J.A. (2004) Biochemical and molecular characterization of a laccase from the edible straw mushroom *Volvariella volvacea*. *Eur. J. Biochem., 271,* 318-328.
[91] Hakala, T.K., Lundell, T., Galkin, S., Maijala, P., Kalkkinen, N., Hatakka, A. (2005) Manganese peroxidases, laccases and oxalic acid from the selective white rot fungus *Physisporinus rivulosus* grown on spruce wood chips. *Enzyme Microbial. Technol., 36,* 461-468.
[92] Buswell, J.A., Odier, E. (1987) Lignin biodegradation. *Crit. Rev. Biotechnol., 6,* 1-60.
[93] Ruttimann-Johnson, C., Cullen, D., Lamar, R.T. (1994) Manganese peroxidases of the white rot fungus *Phanerochaete sordida*. *Appl. Environ. Microbiol., 60,* 599-605.
[94] Kaal, E.E.J., de Jong, E., Field, J.A. (1993) Stimulation of ligninolytic peroxidase activity by nitrogen nutrients in the white rot fungus *Bjerkandera* sp. strain BOS55. *Appl. Environ. Microbiol., 59,* 4031-4036.
[95] Mester, T., Field, J.A. (1997) Optimization of manganese peroxidase production by the white rot fungus *Bjerkandera* sp. Strain BOS55. *FEMS Microbiol. Lett., 155,* 161-168.
[96] Mancilla, R.A., Canessa, P., Manubens, A., Vicuna, R. (2010) Effect of manganese on the secretion of manganese-peroxidase by the basidiomycetes *Ceriporiopsis subvermispora*. *Fungal Gen. Biol., 47,* 656-661.
[97] Shin, K.-S., Kim, Y.H., Lim, J.-S. (2005) Purification and characterization of manganese peroxidase of the white rot fungus *Irpex lacteus*. *J. Microbiol., 43,* 503-509.
[98] Sarkar, S., Martinez, A.T., Martinez, M.J. (1997) Biochemical and molecular characterization of a manganese peroxidase isoenzyme from *Pleurotus ostreatus*. *Biochem. Biophys. Acta., 1339,* 23-30.
[99] Levin, L., Melignani, E., Ramos, A.M. (2010) Effect of nitrogen sources and vitamins on ligninolytic enzyme production by some white rot fungi. Dye decolorization by selected culture filtrates. *Bioresource Technol., 101,* 4554-4563.
[100] Collins, P., Dobson, A.D.W. (1995) Extracellular lignin and manganese peroxidase production by the white rot fungus *Coriolus versicolor* 290. *Biotechnol. Lett., 17,* 989-992.
[101] Jager, A., Croan, S., Kirk, T.K. (1985) Production of ligninases and degradation of lignin in agitated submerged cultures of *Phanerochaete chrysosporium*. *Appl. Environ. Microbiol., 50,* 1274-1278.

[102] Martinez, M.J., Ruiz-Duenas, F.J., Guillen, F., Martinez, A.T. (1996) Purification and catalytic properties of two manganese peroxidase isoenzymes from *Pleurotus eryngii. Eur. J. Biochem., 237,* 424-432.

[103] Mester, T., Field, J.A. (1998) Characterization of a novel manganese peroxidase-lignin peroxidase hybrid isozyme produced by *Bjerkandera* species strain BOS55 in the absence of manganese. *J. Biol. Chem., 273,* 15412-15417.

[104] Brown, J.A., Glenn, J.K., Gold, M.H. (1990) Manganese regulates expression of manganese peroxidase by *Phanerochaete chrysosporium. J. Bacteriol., 172,* 3125-3130.

[105] Bonnarme, P., Jeffries, T.W. (1990) Mn^{2+} regulation of lignin peroxidases and manganese-dependent peroxidases from lignin-degrading white rot fungi. *Appl. Environ. Microbiol., 56,* 210-217.

[106] Perie, F.H., Gold, M.H. (1991) Manganese regulation of manganese peroxidase expression and lignin degradation by the white rot fungus *Dichomitus squalens. Appl. Environ. Microbiol., 57,* 2240-2245.

[107] Steffen, K.T., Hofrichter, M., Hatakka, A. (2002) Purification and characterization of manganese peroxidases from the litter-decomposing basidiomycetes *Agrocybe praecox* and *Stropharia coronilla. Enzyme Microbial. Technol., 30,* 550-555.

[108] Rodriguez, E., Nuero, O., Guillen, F., Martinez, A.T., Martinez, M.J. (2004) Degradation of phenolic and non-phenolic aromatic pollutants by four *Pleurotus* species: the role of laccase and versatile peroxidase. *Soil Biol. Biochem., 36,* 909-916.

[109] Faison, B.D., Kirk, T.K., Farrell, R.L. (1986) Role of veratryl alcohol in regulating ligninase activity in *Phanerochaete chrysosporium. Appl. Environ. Microbiol., 52,* 251-254.

[110] Lundell, T., Leonowicz, A., Rogalski, J., Hatakka, A. (1990) Formation and action of lignin-modifying enzymes in cultures of *Phlebia radiata* supplemented with veratric acid. *Appl. Environ. Microbiol., 56,* 2623-2629

[111] Gill, P.K., Arora D.S. (2003) Effect of culture conditions on manganese peroxidases production and activity by some white rot fungi. *J. Ind. Microbiol. Biotechnol., 30,* 28-33.

[112] Schlosser, D., Grey, R., Fritsche, W. (1997) Patterns of ligninolytic enzymes in *Trametes versicolor.* Distribution of extra- and intracellular enzyme activities during cultivation on glucose, wheat straw and beech wood. *Appl. Microbiol. Biotechnol., 47,* 412-418.

[113] Sethuraman, A., Akin, D.E., Eisele, J.G., Eriksson, K.-E. (1998) Effect of aromatic compounds on growth and ligninolytic enzyme production of two white rot fungi *Ceriporiopsis subvermispora* and *Cyathus stercoreus. Can. J. Microbiol., 44,* 872-885.

[114] Camarero, S., Bockle, B., Martinez, M.J., Martinez, A.T. (1996) Manganese-mediated lignin degradation by *Pleurotus pulmonarius. Appl. Environ. Microbiol., 62,* 1070-1072.

[115] Cohen, R., Persky, L., Hazan-Eitan, Z., Yarden, O., Hadar, Y. (2002) Mn^{2+} alters peroxidase profiles and lignin degradation by the white rot fungus *Pleurotus ostreatus* under different nutritional and growth conditions. *Appl. Biochem. Biotechnol., 102-103,* 415-429.

[116] Steffen, K.T., Hofrichter, M., Hatakka, A. (2000) Mineralisation of ^{14}C-labelled synthetic lignin and ligninolytic enzyme activities of litter-decomposing basidiomycetous fungi. *Appl. Microbiol. Biotechnol., 54,* 819-825.

[117] Deguchi, T., Kakezawa, M., Nishida, T. (1997) Nylon biodegradation by lignin-degrading fungi. *Appl. Environ. Microbiol., 63,* 329-331.

[118] Pozdnyakova, N. (2012) Involvement of the ligninolytic system of white rot and litter-decomposing fungi in the degradation of polycyclic aromatic hydrocarbons. Review article. *Biotechnol. Res. Intern., 2012,* Article ID 243217, 20 pages. DOI:10.1155/2012/243217

[119] Johansson, T., Nyman, P.O. (1993) Isozymes of lignin peroxidase and manganese (II) peroxidase from the white rot basidiomycete *Trametes versicolor*. I. Isolation of enzyme forms and characterization of physical and catalytic properties. *Arch. Biochem. Biophys., 300*, 49-56.

[120] Lundell, T., Schoemaker, H., Hatakka, A., Brunow G. (1993) New mechanism of the Cα-Cβ cleavage in non-phenolic arylglycerol β-aryl ether lignin substructures catalyzed by lignin peroxidase. *Holzforschung, 47*, 219-224.

[121] Edwards, S.L, Raag, R., Wariishi, H., Gold, M.H., Poulos, T.L. (1993) Crystal structure of lignin peroxidase. *Proc. Nat. Acad.Sci. USA, 90*, 750-754.

[122] Choinowski, T., Blodig, W., Winterhalter, K. H., Piontek, K. (1999) The crystal structure of lignin peroxidase at 1.70 Å resolution reveals a hydroxy group on the Cβ of tryptophan 171: A novel radical site formed during the redox cycle. *J. Mol. Biol., 286,* 809-827.

[123] Miki, K., Renganathan, V., Gold, M.H. (1986) Mechanism of β-aryl ether dimeric lignin model compound oxidation by lignin peroxidase of *Phanerochaete chrysosporium. Biochemistry, 25*, 4790-4796.

[124] Hammel, K.E., Jensen, K.A., Mozuch, M.D., Landucci, L.L., Tien, M., Pease, E.A. (1993) Ligninolysis by a purified lignin peroxidase. *J. Biol. Chem., 268,* 12274-12281.

[125] Tien, M., Kirk, T.K., Bull, C., Fee, J.A. (1986) Steady-state and transient-state kinetic studies on the oxidation of 3,4-dimethoxybenzyl alcohol catalyzed by the ligninase of *Phanerocheate chrysosporium* Burds. *J Biol Chem., 261,* 1687-1693.

[126] Schoemaker, H.E., Lundell, T., Hatakka, A., Piontek, K. (1994) The oxidation of veratryl alcohol, dimeric lignin models and lignin by lignin peroxidase: the redox cycle revisioned. *FEMS Microbiol. Rev., 13,* 321-332.

[127] Smith, A.T., Doyle, W.A., Dorlet, P., Ivancich, A. (2009) Spectroscopic evidence for an engineered, catalytically active trp radical that creates the unique reactivity of lignin peroxidase. *Proc. Natl. Acad. Sci. USA, 106*, 16084-16089.

[128] Doyle, W.A., Blodig, W., Veitch, N.C., Piontek, K., Smith, A.T. (1998) Two substrate interaction sites in lignin peroxidase revealed by site-directed mutagenesis. *Biochemistry, 37,* 15097-15105.

[129] Hammel, K. E., Cullen, D. (2008) Role of fungal peroxidases in biological ligninolysis. *Curr. Opin. Plant Biol., 1,* 349-355.

[130] Pérez-Boada, M., Ruiz-Dueñas, F.J., Pogni, R., Basosi, R., Choinowski, T., Martínez, M.J., Piontek, K., Martínez, A.T. (2005) Versatile peroxidase oxidation of high redox potential aromatic compounds: site-directed mutagenesis, spectroscopic and crystallographic investigation of three long-range electron transfer pathways. *J. Mol. Biol., 354*, 385-402.

[131] Miki, Y., Calvino, F.R., Pogni, R., Giansanti, S., Ruiz-Duenas, F.J., Martínez, M.J., Basosi, R., Romero, A., Martínez, A.T. (2011) Crystallographic, kinetic, and

spectroscopic study of the first ligninolytic peroxidase presenting a catalytic tyrosine. *J. Biol. Chem., 286,* 15525-15534.

[132] Miki, Y., Tanaka, Nakamura, M., Wariishi, H. (2006) Isolation and characterization of a novel lignin peroxidase from the white rot basidiomycete *Trametes cervina. J. Fac. Agr. Kyushu Univ., 51,* 99-104.

[133] Miki, Y., Ichinose, H., Wariishi, H. (2010) Molecular characterization of lignin peroxidase from the white rot basidiomycete *Trametes cervina*: a novel fungal peroxidase. *FEMS Microb. Lett., 304,* 39-46.

[134] Sasaki, S., Nonaka, D., Wariishi, H., Tsutsumi, Y., Kondo, R. (2008) Role of Tyr residues on the protein surface of cationic cell-wall peroxidase (CWPO-C) from poplar: potentially unique oxidation sites for oxidative polymerization of lignin. *Phytochemistry, 69,* 348-355.

[135] Johijima, T., Wariishi, H., Tanaka, H. (2002) Veratryl alcohol binding sites of lignin peroxidase from *Phanerochaete chrysosporium. J. Mol. Catal. B: Enz., 17,* 49-57.

[136] Kersten, P., Cullen, D. (2007) Extracellular oxidative systems of the lignin-degrading Basidiomycete *Phanerochaete chrysosporium. Fungal Gen. Biol., 44,* 77-87.

[137] Teunissen, P.J.M. and Field, J.A. (1998) 2-Chloro-1,4-dimethoxybenzene as a novel catalytic cofactor for oxidation of anisyl alcohol by lignin peroxidase. *Appl. Environ. Microbiol., 64,* 830-835.

[138] Ander, P., Messner, K. (1998) Oxidation of 1-hydroxybenzotriazole by laccase and lignin peroxidase. *Biotechnol. Technol., 12,* 191–195.

[139] Bacciocchi, E., Gerini, M.F., Lanzalunga, O., Mancinelli, S. (2002) Lignin peroxidase catalyzed oxidation of 4-methoxymandelic acid. The role of mediator structure. *Tetrahedron, 58,* 8087-8093.

[140] Yadav, M., Singh, S.K., Yadava, S. (2012) Purification, characterization and coal depolymerisation activity of lignin peroxidase from *Lenzitus betulina* MTCC-1183. *Appl. Biochem. Microbiol., 48,* 646-652.

[141] Valli, K., Wariishi, H., Gold, M.H. (1990) Oxidation of monomethoxylated aromatic compounds by lignin peroxidase: Role of veratryl alcohol in lignin biodegradation. *Biochemistry, 29,* 8535-8539.

[142] Yoshida, S., Watanabe, T., Honda, Y., Kuwahara, M. (1997) Effects of water-miscible organic solvents on the reaction of lignin peroxidase of *Phanerochaete chrysosporium. J. Mol. Catalysis B: Enzymatic, 2,* 243-251.

[143] Hammel, K.E., Tardone, P.J. (1988) The oxidative 4-dechlorination of polychlorinated phenols is catalyzed by extracellular fungal lignin peroxidase. *Biochemistry, 27,* 6563-6568.

[144] Chung, N., Aust, S. (1995b) Veratryl alcohol-mediated indirect oxidation of pentachlorophenol by lignin peroxidase. *Arch. Biochem. Biophys., 322,* 143-148.

[145] Collins, P.J., Field, J.A., Teunissen, P., Dobson, A.D.W. (1997) Stabilization of lignin peroxidases in white rot fungi by tryptophan. *Appl. Environ. Microbiol., 63,* 2543-2548.

[146] Ward, G., Belinky, P.A., Hadar, Y., Bilkis, I., Dosoretz, C.G. (2002) The influence of non-phenolic mediators and phenolic co-substrates on the oxidation of 4-bromophenol by lignin peroxidase. *Enzyme Microb. Technol., 30,* 490-498.

[147] Baciocchi, E., Fabbri, C., Lanzalunga, O. (2003) Lignin peroxidase catalyzed oxidation of nonphenolic trimeric lignin model compounds: Fragmentation reactions in the intermediate radical cations. *J. Org. Chem., 68,* 9061-9069.

[148] Haemmerli, S.D., Leisola, M.S.A., Sanglard, D., Fiechter, A. (1986) Oxidation of benzo[a]pyrene by extracellular ligninases of *Phanerochaete chrysosporium*. Veratryl alcohol and stability of ligninase. *J. Biol. Chem., 261,* 6900-6903.

[149] Torres, E., Tinoco, R., Vazquez-Duhalt, R. (1997) Biocatalytic oxidation of polycyclic aromatic hydrocarbons in media containing organic solvents. *Water Science Technol., 36,* 37-44.

[150] Field, J.A., Vledder, R.H., van Zelst, J.G., Rulkens, W.H. (1996) The tolerance of lignin peroxidase and manganese dependent peroxidase to miscible solvents and the *in vitro* oxidation of anthracene in solvent: water mixtures. *Enzyme Microbial Technol., 18,* 300-308.

[151] Vazquez-Duhalt, R., Westlake, D.W.S., Fedorak, P.M. (1994) Lignin peroxidase oxidation of aromatic compounds in systems containing organic solvents. *Appl. Environ. Microbiol., 60,* 459-466.

[152] Hammel, K.E., Kalyanaraman, B., Kirk, T.K. (1986) Oxidation of polycyclic aromatic hydrocarbons and dibenzo[p]dioxins by *Phanerochaete chrysosporium* ligninase. *J. Biol. Chem., 261,* 16948-16952.

[153] Khindaria, A., Grover, T.A., Aust, S.D. (1995) Reductive dehalogenation of aliphatic halocarbons by lignin peroxidase of *Phanerochaete chrysosporium. Environ. Sci. Technol., 29,* 719-725.

[154] Hammel, K.E., Moen, M.A. (1991) Depolymerization of a synthetic lignin in vitro by lignin peroxidase. *Enzyme Microb. Technol., 13*, 15-18.

[155] Joshi, D.K., Gold, M.H. (1993) Degradation of 2,4,5-trichlorophenol by the lignin-degrading basidiomycete *Phanerochaete chrysosporium. Appl. Environ. Microbiol., 59,* 1779-1785.

[156] Ruiz-Duenas, F.J., Morales, M., Garcia, E., Miki, Y., Martinez, M.J., Martinez, A.T. (2008) Substrate oxidation sites in versatile peroxidase and other basidiomycetes peroxidase. *J. Exp. Bot., 60,* 441-452.

[157] Michels, J., Gottschalk, G. (1994) Inhibition of the lignin peroxidase of *Phanerochaete chrysosporium* by hydroxylamino-dinitrotoluene, an early intermediate in the degradation of 2,4,6-trinitrotoluene. *Appl. Environ. Microbiol., 60*, 187-194.

[158] Podgornik, H., Grgic, I., Perdith, A. (1999) Decolorization rate of dyes using lignin peroxidase of *Phanerochaete chrysosporium. Chemosphere, 38,* 1353-1359.

[159] Ollika, P., Alhonmaki, K., Leppanen, V., Glumoff, T., Raijola, T., Suominen, I. (1993) Decolorization of azo, triphenylmethane, heterocyclic and polymeric dyes by lignin peroxidase izoenzymes from *Phanerochaete chrysosporium. Appl. Environ. Microbiol.* 59, 4010-4016.

[160] Ferreira, V.S., Magalhaes, D.B., Kling, S.H., Da Dilva, JR J.G., Bon, E.P.S. (2000) *N*-Demethylation of methylene blue by lignin peroxidase from *Phanerochaete chrysosporium*. Stoichiometric relation for H_2O_2 consumption. *Appl. Biochem. Biotechnol., 84-86*, 255-256.

[161] Manimekalai, R., Swaminathan, T. (2000) Removal of hazardous compounds by lignin peroxidase from *Phanerochaete chrysosporium. Bioproc. Engin., 22,* 29-33.

[162] Woo, S.H., Cho, J.S., Lee, B.S., Kim, E.K. (2004) Decolorization of melanin by lignin peroxidase from *Phanerochaete chrysosporium. Biotechnol. Bioprocess. Engin., 9,* 256-260.

[163] Ralph, J.P., Catcheside, D.E.A. (1999) Transformation of macromolecules from a brown coal by lignin peroxidase. *Appl. Microbiol. Biotechnol.*, *52*, 70-77.
[164] Hirai, H., Sugiura, M., Kawai, S., Nishida, T. (2005) Characterization of novel lignin peroxidase produced by white rot fungus *Phanerochaete sordida* YK-624. *FEMS Microbiol. Lett.*, *246*, 19-24.
[165] Leisola, M.S.A., Haemmerli, S.D., Waldner, R., Schoemaker, H.E., Schmidt, H.W.H., Fiechter, A. (1988) Metabolism of a lignin model compound, 3,4-dimethoxybenzyl alcohol by *Phanerochaete chrysosporium*. *Cellulose Chem. Technol.*, *22*, 267-277.
[166] Mester, T, Ambert-Balay, K, Ciofi-Baffoni, S, Banci, L, Jones, AD, Tien, M. (2001) Oxidation of a tetrameric nonphenolic lignin model compound by lignin peroxidase. *J. Biol. Chem.*, *276*, 22985-22990.
[167] Gilardi, G., Harvey, P.J., Cass, A.E.G., Palmer, J.M. (1990) Radical intermediates in veratryl alcohol oxidation by ligninase – NMR evidence. *Biochim. Biophys. Acta*, *1041*, 129-132.
[168] Cai, D, Tien, M. (1992) Kinetic studies on the formation and decomposition of compounds II and III – reactions of lignin peroxidase with H_2O_2. *J. Biol. Chem.*, *267*, 11149-11155.
[169] Lundell, T., Hatakka, A. (1994) Participation of Mn (II) in the catalysis of laccase, manganese peroxidase and lignin peroxidase from *Phlebia radiata*. *FEBS Lett.*, *348*, 291-296.
[170] Valderrama, B., Ayala, M., Vazquez-Duhalt, R. (2002) Suicide inactivation of peroxidases and the challenge of engineering more robust enzymes. *Chem. Biol.*, *9*, 555-565.
[171] Mester, T., Tien, M. (2000) Oxidation mechanism of ligninolytic enzymes involved in the degradation of environmental pollutants. *Int. Biodeter. Biodegr.*, *46*, 51-59.
[172] Palma, C., Martinez, A. T., Lema, J. M., Martinez, M. J. (2000) Different fungal manganese-oxidizing peroxidases: a comparison between *Bjerkandera* sp. and *Phanerochaete chrysosporium*. *J. Biotechnol.*, *77*, 235-245.
[173] Boer, C., Obici, L., Desouza, C., Peralta, R. (2006) Purification and some properties of Mn-peroxidase from *Lentinula edodes*. *Process Biochem.*, *41*, 1203-1207.
[174] Cheng, X., Jia, R., Li, P., Tu, S., Zhu, Q., Tang, W., Li, X. (2007) Purification of a new manganese peroxidase of the white rot fungus *Schizophyllum* sp. F17, and decolorization of azo dyes by the enzyme. *Enzyme Microb. Tech.*, *41*, 258-264.
[175] Hoshino, F., Kajino, T., Sugiyama, H., Asami, O., Takahashi, H. (2002) Thermally stable and hydrogen peroxide tolerant manganese peroxidase (MnP) from *Lenzites betulinus*. *FEBS Lett.*, *530*, 249-252.
[176] Bermek, H., Yazıcı, H., Öztürk, H., Tamerler, C., Jung, H., Li, K., Brown, K.M., Ding, H., Xu, F. (2004) Purification and characterization of manganese peroxidase from wood-degrading fungus *Trichophyton rubrum* LSK-27. *Enzyme Microbial. Technol.*, *35*, 87-92.
[177] Banci, L. (1997) Structural properties of peroxidases. *J. Biotechnol.*, *53*, 253-263.
[178] Conesa, A., Punt, P.J., van den Hondel, C.A.M.J.J. (2002) Fungal peroxidases: molecular aspects and applications. *J. Biotechnol.*, *93*, 143-158.
[179] Sundaramoorthy, M., Gold, M.H., Poulos, T.L. (2010) Ultrahigh (0.93 Å) resolution structure of manganese peroxidase from *Phanerochaete chrysosporium*: implications for the catalytic mechanism. *J. Inorg. Bioch.*, *104*, 683-690.

[180] Parveen, K., Usha, K.Y., Visvanath, B., Reddy, B.R. (2012) Kinetic properties of manganese peroxidase from the mushroom *Stereum ostrea* and its ability to decolorize dyes. *J. Microbiol. Biotechnol. 22,* 1540-1548.

[181] Petruccioli, M., Frasconi, M., Quaratino, D., Covino, S., Favero, G., Mazzei, F., Federici, F., D'Annibale, A. (2009) Kinetic and redox properties of MnP II, a major manganese peroxidase isoenzyme from *Panus tigrinus* CBS 577.792009. *J. Biol. Inorg. Chem., 14,* 1153-1163.

[182] Singh, S.K., Yadav, M., Yadava, S., Yadav, K.D.S. (2011) Coal depolimerising activity and haloperoxidase activity of Mn-peroxidase from *Fomes durissimus* MTCC-1173. *Bioinorg. Chem. Appl., 2011,* ID 260802, 8 pages. DOI: 10.1155/2011/260802.

[183] Kuan, I.C., Johnson, K.J., Tien, M. (1993) Kinetic analysis of manganese peroxidase. The reaction with manganese complexes. *J. Biol. Chem., 268,* 20064-20070.

[184] Xiaobin, C., Pingsheng, L., Shiqian, T., Qin, Z., Wenzhong, T., Xudong, L. (2007) Purification of a new manganese peroxidase of the white rot fungus *Schizophyllum* sp. F17, and decolorization of azo dyes by enzyme. *Enzyme Microbial Technol., 41,* 258-264.

[185] Matsubara, M., Suzuki, J., Deguchi, T., Miura, M., Kitaoka, Y. (1996) Characterization of manganese peroxidases from the hyperligninolytic fungus IZU-154. *Appl. Environ. Microbiol., 62,* 4066-4072.

[186] Asher, M., Iqbal, H.M.N. (2011) Characterization of a novel manganese peroxidase purified from solid state culture of *Trametes versicolor* IBL-04. *BioResources, 6,* 4302-4315.

[187] Sack, U., Hofrichter, M., Fritsche, W. (1997) Degradation of polycyclic aromatic hydrocarbons by manganese peroxidase of *Nematoloma frowardii. FEMS Microbiol. Lett., 152,* 227-234.

[188] Eibes, G., Lu-Chau, T., Feijoo, G., Moreira, M.T., Lema, J.M. (2005) Complete degradation of anthracene by manganese peroxidase in organic solvent mixtures. *Enzyme Microbial. Technol., 37,* 365-372.

[189] Baborová, P., Möder, M., Baldrian, P., Cajthamlová, K., Cajthaml, T. (2006) Purification of a new manganese peroxidase of the white rot fungus *Irpex lacteus*, and degradation of polycyclic aromatic hydrocarbons by the enzyme. *Res Microbiol, 157,* 248-253.

[190] Acevedo, F., Pizzul, L., Castillo, M., Cuevas, R., Diez, M. (2011) Degradation of polycyclic aromatic hydrocarbons by the Chilean white rot fungus *Anthracophyllum discolor. J. Hazard. Mater., 185,* 212-219.

[191] Moen, M.A., Hammel, K.E. (1994) Lipid peroxidation by the manganese peroxidase of *Phanerochaete chrysosporium* is the basis for phenanthrene oxidation by the intact fungus. *Appl. Environ. Microbiol., 60,* 1956-1961.

[192] Wariishi, H., Valli, K., Gold, M.H. (1989) Oxidative cleavage of a phenolic diarylpropane lignin model dimer by manganese peroxidase from *Phanerochaete chrysosporium. Biochemistry, 28,* 6017-6023.

[193] Urzua, U., Kersten, P.J., Vicuña R. (1998) Manganese peroxidase-dependent oxidation of glyoxylic acid synthetized by *Ceriporiopsis subvermispora* produces extracellular hydrogen peroxide. *Appl. Environ. Microbiol., 64,* 68-73.

[194] Scheibner, K., Hofrichter, M., Fritsche, W. (1997) Mineralization of 2-amino-4,6-dinitrotoluene by manganese peroxidase of the white rot fungus *Nematoloma frowardii*. *Biotechnol. Lett., 19,* 835-839.

[195] Bao, W., Fukushima, Y., Jensen, K.A., Moen, M.A., Hammel, K.E. (1994) Oxidative degradation of non-phenolic lignin during lipid peroxidation by fungal manganese peroxidase. *FEBS Lett., 354,* 297-300.

[196] Eibes, G., Cajthaml, T., Moreira, M.T., Feijoo, G., Lema, J.M. (2006) Enzymatic degradation of anthracene, dibenzothiophene and pyrene by manganese peroxidase in media containing acetone. *Chemosphere, 64,* 408-414.

[197] Günther, T., Sack, U., Hofrichter, M., Latz, M. (1998) Oxidation of PAH and PAH-derivatives by fungal and plant oxidoreductases. *J. Basic Microbiol., 38,* 113-122.

[198] Moldes, D., Rodríguez Couto, S., Cameselle, C., Sanromán, M. A. (2003) Study of the degradation of dyes by MnP of *Phanerochaete chrysosporium* produced in a fixed-bed bioreactor. *Chemosphere, 51,* 295-303.

[199] Novotný, Č., Svobodová, K., Erbanová, P., Cajthaml, T., Kasinath, A., Lang, E., Šašek, V. (2004) Ligninolytic fungi in bioremediation: extracellular enzyme production and degradation rate. *Soil Biol. Biochem., 36,* 1545-1551.

[200] Kokol, V., Doliška, A., Eichlerová, I., Baldrian, P., Nerud, F. (2007) Decolorization of textile dyes by whole cultures of *Ischnoderma resinosum* and by purified laccase and Mn-peroxidase. *Enzyme Microb. Tech., 40,* 1673-1677.

[201] Wang, J., Ogata, M., Hirai, H., Kawagishi, H. (2011) Detoxification of aflatoxin B_1 by manganese peroxidase from the white rot fungus *Phanerochaete sordida* YK-624. *FEMS Microbiol. Lett., 314,* 164-169.

[202] Hofrichter, M., Scheibner, K., Schneegas, I., Fritsche, W. (1998) Enzymatic combustion of aromatic and aliphatic compounds by manganese peroxidase from *Nematoloma frowardii. Appl. Environ. Microbiol., 64,* 399-404.

[203] Kapich, A., Hofrichter, M., Vares, T., Hatakka, A. (1999) Coupling of manganese peroxidase-mediated lipid peroxidation with destruction of nonphenolic lignin model compounds and ^{14}C-labeled lignins. *Biochem. Biophys. Res. Commun., 259,* 212-219.

[204] Hofrichter, M., Ziegenhagen, D., Vares, T., Friedrich, M., Jager, M.G., Fritsche, W., Hatakka, A. (1998) Oxidative decomposition of malonic acid as basis for the action of manganese peroxidase in the absence of hydrogen peroxide. *FEBS Lett., 434,* 362-366.

[205] Kanayama, N., Suzuki, T., Kawai, K. (2002) Purification and characterization of an alkaline manganese peroxidase from *Aspergillus terreus* LD-1. *J. Biosci. Bioeng., 93,* 405-410.

[206] Rodakiewicz-Nowak, J., Jarosz-Wilkolazka, A., Luterek, J. (2006) Catalytic activity of versatile peroxidase from *Bjerkandera fumosa* in aqueous solutions of water–miscible organic solvents. *Appl. Catalysis A: General, 308,* 56-61.

[207] Moreira, P.R., Almeida-Vara, E., Malcata, F.X., Duarte, J.C. (2007) Lignin transformation by a versatile peroxidase from a novel *Bjerkandera* sp. strain. *Int. Biodeterior. Biodegrad., 59,* 234-238.

[208] Heinfling, A., Martinez, M.J., Martinez, A.T., Bergbauer, M., Szewzyk, U. (1998) Purification and characterization of peroxidases from dye-decolorizing fungus *Bjerkandera adusta. FEMS Microbiol. Lett., 165,* 43-50.

[209] Chen, M., Yao, S., Zhang, H., Liang, X. (2010) Purification and characterization of a versatile peroxidase from edible mushroom *Pleurotus eryngii. Chinese J. of Chem. Eng., 18,* 824-829.

[210] Tinoco, R., Verdin, J., Vazquez-Duhalt, R. (2007) Role of oxidizing mediators and tryptophan 172 in the decoloration of industrial dyes by the versatile peroxidase from *Bjerkandera adusta. J. Mol. Catalysis B: Enzymatic, 46,* 1-7.

[211] Garcia-Ruiz, E., Gonzalez-Perez, D., Ruiz-Dueñas, F.J., Martínez, A.T., Alcalde, M. (2012) Directed evolution of a temperature-, peroxide- and alkaline pH-tolerant versatile peroxidase. *Biochem. J., 441,* 487-498.

[212] Taboada-Puig, R., Lu-Chau, T., Moreira, M.T., Feijoo, G., Martinez, M.J., Lema, J.M. (2011) A new strain of *Bjerkandera* sp. production, purification and characterization of versatile peroxidase. *World J. Microbiol. Biotechnol., 27,* 115-122.

[213] Hofrichter, M., Ullrich, R., Pecyna, M.J., Liers, C., Lundell, T. (2010) New and classic families of secreted fungal heme peroxidases. *Appl. Microbiol. Biotechnol., 87,* 871-897.

[214] Davila-Vazquez, G., Tinoca, R., Pickard, M.A., Vazquez-Duhalt, R. (2005) Transformation of halogenated pesticides by versatile peroxidase from *Bjerkandera adusta. Enzyme Microbial Technol., 36,* 223-231.

[215] Eibes, G., Debernardi, G., Feijoo, G., Moreira, T., Lema, J. (2011) Oxidation of pharmaceutically active compounds by a ligninolytic fungal peroxidase. *Biodegradation, 22,* 539-550.

[216] Pozdnyakova, N., Makarov, O., Chernyshova, M., Turkovskaya, O., Jarosz-Wilkolazka, A. (2013) Versatile peroxidase of *Bjerkandera fumosa*: Substrate and inhibitor specificity. *Enzyme Microbial Technol., 52,* 44-53.

[217] Wang, Y., Vazquez-Duhalt, R., Pickard, M.A. (2003) Manganese-lignin peroxidase hybrid from *Bjerkandera adusta* oxidizes polycyclic aromatic hydrocarbons more actively in the absence of manganese. *Can. J. Microbiol., 49,* 675-682.

[218] Morales, M., Mate, M.J., Romero, A., Martinez, M.J., Martinez, A.T. (2012) Two oxidation sites for low redox-potential substrates: A directed mutagenesis, kinetic and crystallographic study on *Pleurotus eryngii* versatile peroxidase. *J. Biol. Chem., 287,* 41053-41067.

[219] Camarero, S., Sarkar, S., Ruiz-Dueñas, F.J., Martínez, M.J., Martínez, T. (1999) Description of a versatile peroxidase involved in the natural degradation of lignin that has both manganese peroxidase and lignin peroxidase substrate interaction sites. *J. Biol. Chem., 274,* 10324-10330.

[220] Wang, Y., Vazquez-Duhalt, R., Pickard, M. (2002) Purification, characterisation, and chemical modification of manganese peroxidase from *Bjerkandera adusta* UAMH 8258. *Curr. Microbiol., 45,* 77-87.

[221] Heinfling, A., Martinez, M.J., Martinez, A.T., Bergbauer, M., Szewzyk, U. (1998) Transformation of industrial dyes by manganese peroxidase from *Bjerkandera adusta* and *Pleurotus eryngii* in a manganese-independent reaction. *Appl. Environ. Microbiol., 64,* 2788-2793.

[222] Gomez-Toribio, V., Martinez, A.T., Martinez, M.J., Guillen, F. (2001) Oxidation of hydroquinones by the versatile peroxidase from *Pleurotus eryngii. Eur. J. Biochem., 268,* 4787-4793.

[223] Taboada-Puig, R, Lú-Chau, T.A., Moreira, M.T., Feijoo, G., Lema, J.M. (2013) Activation of kraft lignin by an enzymatic treatment with a versatile peroxidase from *Bjerkandera* Sp. R1. *Appl. Biochem. Biotechnol. 169,* 1262-1278.

[224] Moreira, M.T., Sierra-Alvarez, R., Lema, J.M., Feijo, G., Field, J.A. (2001) Oxidation of lignin in eucalyptus kraft pulp by manganese peroxidase from *Bjerkandera* sp. strain BOS55. *Bioresource Technol., 78,* 71-79.

[225] Tsukihara, T., Honda, Y., Sakai, R., Watanabe, T., Watanabe, T. (2008) Mechanism for oxidation of high-molecular-weight substrates by a fungal versatile peroxidase, MnP2. *Appl. Environ. Microbiol., 74,* 2873-2881.

[226] Hibi, M., Hatahira, S., Nakatani, M., Yokozeki, K., Shimizu, S., Ogawa, J. (2012) Extracellular oxidases of *Cerrena* sp. complementarily functioning in artificial dye decolorization including laccase, manganese peroxidase, and novel versatile peroxidases. *Biocatal. Agricul. Biotechnol., 1,* 220-225.

[227] Verdín, J., Pogni, R., Baeza, A., Baratto, M.C., Basosi, R., Vázquez-Duhalt, R. (2006) Mechanism of versatile peroxidase inactivation by Ca^{2+} depletion. *Biophys. Chem. 12,* 163-70.

[228] Hofrichter, M., Lundell, T., Hatakka, A. (2001) Conversion of milled pine wood by manganese peroxidase from *Phlebia radiata. Appl. Environ. Microbiol., 67,* 4588-4593.

[229] Pinto, P.A., Dias, A.A., Fraga, I., Marques, G., Rodrigues, M.A.M., Colac0, J., Sampario, A., Bezerra, R.M.F. (2012) Influence of ligninolytic enzymes on straw saccharification during fungal pretreatment. *Biores. Technol., 111,* 261-267.

[230] Roessl, U., Nahalka, J., Nidetzky, B. (2010) Carrier-free immobilized enzymes for biocatalysts. *Biotechnol. Lett., 32,* 341-350.

[231] Ferrer, I., Dezotti, M., Duran, N. (1991) Decolorization of Kraft effluent by free and immobilized lignin peroxidases and horseradish peroxidase. *Biotechnol. Lett., 13,* 577-582.

[232] Asgher, M., Iqbal, H.M.N., Irshad, M. (2012) Characterization of purified and xerogel immobilized novel lignin peroxidase produced from *Trametes versicolor* IBL-04 using solid state medium of corncobs. *BMC Biotechnology*, *12,* 46-53.

[233] Gianfreda, L., Rao M.A. (2004) Potential of extracellular enzymes in remediation of polluted soils: a review. *Enzyme Microbial Technol., 35,* 339-354.

[234] Nicole, M., Chamberland, H., Rioux, D., Lecours, N., Rio, B., Geiger, J.P., Ouellette, G.B. (1993) A cytochemical study of extracellular sheaths associated with *Rigidoporus lignosus* during wood decay. *Appl. Environ. Microbiol., 59,* 2578-2588.

[235] Cameron, M.D, Timofeevsky, S., Aust S. (2000) Enzymology of *Phanerochaete chrysosporium* with respect to the degradation of recalcitrant compounds and xenobiotics. *Appl. Microbiol. Biotechnol., 54,* 751-758.

In: Peroxidases
Editors: Lucas Bogaert and Noah Coppens

ISBN: 978-1-62808-261-6

Chapter II

Plant Peroxidases: Functions and Regulation during Pathogenesis

A. V. Sorokan' and I. V. Maksimov
Institute of Biochemistry and Genetics, UFA Scientific Centre,
Russian Academy of Science, Russia

Abstract

Peroxidase (donor: H_2O_2oxidoreductase, EC 1.11.1.7) (PO) belongs to a class of widespread and vital enzymes. One of its specific features is the multiplicity of its molecular isoforms exhibiting various functions], and the highly variable genes encoding them are controlled by various *cis*-elements and *trans*-factors that are responsible for their differentiated expression activity depending on environmental conditions and stages of ontogenesis. POs are localizing in cytoplasm, plasma membrane or secrete outside of cell. In cell wall POs presence in the ionic-bound or covalent bound fraction of proteins, and even freely circulated in apoplast. POs are key enzymes of lignification capable of binding with polysaccharides of plant and (or) pathogens. An important physiological sense of these reactions is the strengthening of the cell wall and construction on the pathogens' paths of physical barriers consisting of lignin- and suberin-containing polymers, which also possess high toxicity. This defense reaction is developing more or less automatically, thus these barriers are formed just in the zone of pathogen penetration and concurrently with the beginning of the active pathogen expansion into the tissues of the host plant. However, mechanisms of development of these events are still unclear. Expression of some genes encoding different isoPOs are regulated by pathogens (and their metabolites), elicitors, and stress-related hormones and hormone-like compounds, such as salicylic and jasmonic acids, ethylene and others.

Peroxidases in Plant Defense against Pathogens

One of the amazing properties of all living organisms on the Earth is their ability to defence against pathogens and parasites. Defence system responsible for this has a great spectrum of mechanisms. This mechanisms and their regulation are studied very intensively. Recently, the discovery of the molecular mechanisms of reception of pathogen attack and invasion, transduction signal in the genome of host and triggering of components of host immune system should be considered as an important event in biology. Significant progress in description of the humans and animals immunity was obtained. Finally, it is really important that the immune system of plants, which, like other organisms, need to protect themselves against pathogens and herbivores was declared [1]. When a plant is attacked by pathogens, their defense system is activated. Recently, it has been showed the synthesis *de novo* of defense proteins under the hormonal control, in particular, of salicylic (SA), jasmonic (JA), abscisic (ABA) acids and ethylene in the plant [1, 2-5]. Induction of plants signaling system by pathogens and elicitors leads to production of ROS and nitrogen oxide, antimicrobial metabolites (phytoalexins), strengthening of the cell wall due to its lignification actively contributed by PO (PO) [6]. This enzyme occupies a crucial position in plant defense against pathogens [1, 7]. A direct biocidal effect of products of the PO enzymatic catalysis was shown for virtually all living beings. Thus, in humans and other animals there are several POs, and among them myeloPO or lactoPO have strong antimicrobial efficiency [8].

PO is activated in plants in response to infection both due to constitutive synthesis and to expression of pathogen-inducible isoforms [9]. Activation of PO during pathogenesis correlates with formation in the infection zone of a spontaneous local hypersensitive reaction of plant cells surrounded by a zone impermeable for water and nutrients and resulting in suppression of pathogens growth and development. This reaction can be induced not only by pathogens, but also by hydrolases produced by them. Consequently, this reaction can be initiated by polysaccharides possessing features of elicitors and also by some stress hormones [2-5, 10, 11].

In *Aegilops umbellulata* (42 specimens) the resistance to *S. nodorum* negatively correlated with the PO activity in the cell walls protein fraction [12], and this corresponded to hypotheses about the importance of POs located in the cell wall for the plant resistance to pathogens [13]. In the isoenzyme spectrum of uninfected leaves of resistant forms of Aegilops, a PO isoenzyme was detected with p*I*~ 7.5 [12]. PO with the same p*I*displayed antifungal properties against *Fusarium*-induced dry rot of roots [14].

Purified isoPOs from the *Acorus calamus* AcPOX displayed a high fungicidal activity to *Fusarium moniliforme*, *Macrophomina phaseolina*, *Trichosporium vesiculosum* [15], those from cotton were active against *Verticillium dahlia* [16], and isoPOs from beans were active against *Coprinus comatus*, *Mycosphaerella arachidicola*, *Fusarium oxysporum*, and *Botrytis cinerea* [14]. It should be noted that the high antifungal active PO from sedge cane revealed homology with PO from rice BAD61674, BAD61677, and BAD61671 expressed under the influence of bacteria [15]. The antifungal activity pronouncedly depended on the enzymatic activity. Similar antifungal activities were found in POs from cotton, which were bound with chitin of the pathogen cell wall [16]. Note that in the presence of H_2O_2 the myeloPO of animals inhibited the growth of pathogenic fungi, including the phytopathogens *Fusarium solani*, *Colletotrichum gloeosporioides*, and *C. malvarum* [18]. Therefore, both plant and animal POs can be considered not only as mediators in the plant defence associated with synthesis of ROS, lignin, and suberin, but also as enzymes with direct fungicidal activity.

The influence of POs on systems responsible for resistance against pathogens is specific. Thus, transfer of the pepper ascorbate PO gene *CAPOA1* into tobacco plants resulted in its hyperactivation and induced resistance to the oomycetes *Ralstonia solanacearum* but not to the pathogenic bacterium *Pseudomonas siringa*pv. *tabaci* [19].

The production of H_2O_2 and formation of lignin are contributed both by alkaline and by acidic isoPOs located in the cell walls, and mainly the covalently bound fractions [20]. This is also confirmed by increased activities of other enzymes that are involved in modification of the plant cell walls – phytoalexins and polyphenol oxidases, which promote a rapid accumulation of primary compounds involved in polymerization of lignin with POs [21]. This provides conditions for effective and rational oxidation of phenolic compounds into lignin or suberin polymers and also for formation of phytoalexins [22].

Infection of seedlings of a resistant wheat cultivar with agents of septoriosis, root rots, and smut led to activation of PO isoforms with p*I*~3.5, ~4.5, and ~9.8 that interacted with chitin and glucan from pathogenic fungi [23, 24]. Moreover, on infection with fungal pathogens in resistant plants and also under the influence of elicitors anionic POs were also activated in the enzyme fraction bound with the cell walls [25]. Therefore, some pathways can be suggested for regulation by elicitors of defense reactions in wheat plants infected with fungal pathogens. Thus, chitin induces a defense in plants, first by increasing the expression of anionic PO, and, second by inducing the translocation of isoPOs into the region of cell walls where they are involved in strengthening intermolecular bonds and lignification [25, 26] that later promotes a decrease in the damaging effect of pathogens. Thus, PO is likely to be a component of the active system of plant defense. The importance of POs in regulation of physiological characteristics of plants during their interaction with microorganisms is also observed in bean–rhizobial symbiosis [26, 27]. The plants are supposed to switch on regulatory mechanisms directed to limitation of further penetration of rhizobia and their multiplication or propagation through the tissues by changing the PO activities, which allows the optimal quantity of functional nodules to be formed [27, 28].

Infection of the *Arachis hypogaea* leaves with *P. syringae* induces in plants a short-term expression of the *Ep5C* gene, the product of which resembles an extracellular cationic PO of horseradish and does not react to treatment with salicylic acid, ethylene, methyl jasmonate, and wounding [29]. Recently the *CaPrx02* gene was detected in *Capsicum annum*, whose product regulates the H_2O_2 level during formation the defense response to pathogen invasion [30]. Infection of leaves of the rice *Xanthomonas oryzae* was associated with thickening of the cell wall, which prevented penetration of the pathogen at the point of infection [31]. The cell wall thickening correlated with the activity of *OsPrx114* PO.

Thus, on studies of the plant resistance attention is accentuated on the importance of molecular multiplicity of enzymes for increasing the mobility of adaptation mechanisms and eliminating pathological consequences of stress factors. Certainly, the existence of isoenzymes enhances metabolic potentials of organisms exposed to stress factors, increases the plasticity of the cell metabolism and the capability of supporting the autoregulation system, and induced pathways for optimization of physiological processes disturbed during pathogenesis. Data on the direct antipathogenic influence of POs, including the constitutively synthesized ones, suggest that the enzyme could be an important constituent of the defense system that is efficiently functioning in the initial stages of the organism's defense against pathogens and during the immunity induced by them.

Involvement of Peroxidases in Regulation of Plant Defense

Determination of the role of low molecular weight compounds possessing hormonal and immunomodulating properties in the interaction between the host and pathogens is now an actively developed trend in modern physiology and biochemistry of plants. The hormonal system is one of the most effective systems for regulation of various physiological and biochemical reactions in plants. Correction of hormonal balance in plants and of their total biochemical status by using as modifiers biologically active natural compounds with hormonal functions and even of microorganisms as producers of such compounds is a current strategy in plant defense [4, 28]. Seven PO genes in *A. thaliana* were expressed under the influence of *Pseudomonas* ssp., three genes in the leaves infected with fungi, and two genes on infection by insects [32-34]. Three genes (*AtPrx21*, *AtPrx62*, *AtPrx71*) were induced upon infection with fungi [32], on wounding [35], and also in response to other stresses, which suggests their involvement in the main mechanisms of plant defense. The expression of the *AtPrx62* gene caused an 80% increase in the resistance of transgenic plants, and of the *AtPrx21* and *AtPrx71* caused a 30% increase [36]. The *AtPrx21* gene product is supposed to play a protective role on exposure to pathogens and cytodifferentiation of plants [37]. The *AtPrx71* gene is expressed during insufficiency of phosphorus [38] or oxygen [39], formation of secondary cell wall of vessels [40], under conditions of hyposmolality in cell culture, and increase in the level of H_2O_2 [41]. Wounding resulted in a pronounced activation of cationic and anionic isoPOs in seedlings of *Prosopsis* plants [42]. The activity of anionic POs in plants retained for a longer time than the activity of cationic POs in the presence of an inhibitor of mRNA polyadenylation (cordycepin). These findings suggest that either anionic POs are more stable than cationic ones or that plants have a certain reserve of mRNA.

The ability of the *shpx6b* gene from the *Stylosanthes humilis* to express on infection with the pathogen *C. gloeosporioides* and under the influence of methyl jasmonate and wounding can be used as an effective test system for determination SAR of plants. Being inserted into the tobacco genome, this gene promoted in plants tolerance to *Ph. parasitica* var. *nicotianae*, whereas tobacco plants hypersynthesis of anionic POs displayed an increased biosynthesis of lignin and resistance to *Fusarium oxysporum* and to insects [43].

Regulating of Peroxidase Activity by Phytohormones

In the plant POs regulate levels and activities of some endogenous and exogenous signaling molecules, including those that possess growth-regulating properties [42, 44, 45]. This is realized due to enzyme synthesis *de novo* and releasing of bound forms and also due to enzyme modification by proteases [46-48]. Thus, indolylacetic acid (IAA) is poorly oxidized in the transport system, but the cell wall and apoplast where POs are located are main sites of IAA activity [35]. The ability of POs to oxidize IAA [49, 50] allows them to regulate the amount of the auxin in plant tissues [51] and influence some processes within the cell by changing of plant metabolism from cell growth to cell differentiation. IAA can be oxidized in plants through different pathways, and this is a free radical process. The PO functions as an IAA oxidase depends on the hormone location within the cell [46].

The important role of POs in the regulation of IAA level is also confirmed by the enzyme possessing sites which specific bind IAA [50]. Note that in the molecule of wheat anionic PO the domain adjacent to this site was capable of binding chitin. This was proved by inhibiting the IAA oxidase activity of the enzyme on addition of acetylated chitin into the preincubation or reaction medium of oligosaccharides [52]. These data allow us to consider otherwise the structural–functional organization of POs involved, on one hand, in the regulation of auxin levels and, on the other hand, capability of binding with poly- or oligosaccharides of the pathogen or host cell wall. In addition to these findings, which suggest the involvement of PO in IAA catabolism, there are data on the contribution of PO to catalysis of oxidative decarboxylation of amino acids and to tryptophane conversion into IAA. The interesting observation of the involvement of PO in polymerization or depolymerization of phenolic compounds should also be mentioned.

Substantial data were obtained on mutant tomato plants with low level of abscisic acid (ABA), which were poorly colonized with arbuscular mycorrhiza, and the treatment with ABA promoted infecting the roots with fungi and bacteria [53]. The authors consider this reaction of the plants to be associated with a negative influence of the hormone and pathogen on the PO activity. Note that the influence of ABA on the PO activity depended on both the dose and time of exposure. Although a short-term treatment with ABA increased the enzyme activity, long-term treatment with high concentrations decreased the enzyme activity [53, 54].

As far as PO is important for development of resistance under conditions of both abiotic and biotic stress, the multiplicity of PO forms should be taken into account because this multiplicity suggests a differentiated activation/suppression of individual isoPOs in which expression, in particular, of anionic isoPOs is directly controlled by ABA [55]. However, POs can efficiently function also in the absence of ABA. Thus, there is an ABA-independent pathway of activation of ascorbate PO in cotton plants. On the fluridone-induced decrease in the activities of SOD, catalase, and glutathione reductase, the treatment of calluses with paraquat induced in the cells an ABA-independent activation of PO [55].

The nature of the enzyme catalyzing the last stage of ethylene biosynthesis is still under discussion. The involvement of PO in this process is supposed because production of ethylene correlates with PO activity in plant tissues. In a model system PO can *in vitro* convert 1-aminocyclopropane-1-carbonic acid (ACPC) into ethylene. However, because low specificity of the model system was supposed to generate ROS, which in turn nonenzymatically cleave ACPC with production of ethylene.

Pathogens and their metabolites, elicitors, and hormone-like compounds are effective regulators of the expression of genes encoding the class III isoPOs. Were shown activities of POs and of its genes are induced in the presence of pathogenic fungi [56], bacteria [57], and viruses [58]. Numerous data indicate there is differentiated transcription of genes controlling biosynthesis of certain PO isoenzymes in some tissues and organs or under the influence of various physical, chemical, and also biological stress factors that is associated with the presence of multiple *cis*-elements in the promoter part of the genes encoding this enzyme [37, 59, 46, 55, 60,61].

However, the mechanism of interaction of the transcription regulating factors with DNA-dependent RNA polymerases and promoter regions of the PO-encoding genes is still a poorly studied problem in plant genome functioning. Information concerning the regulation of POs genes expression is especially scarce. Only for 36-44 of 74 genes encoding POs of *A. thaliana* are products of transcriptional activity determined as individual isoPOs [37]. Eight genes with nonspecific and relatively high level of expression are contained in the promoter zone *cis*-elements sensitive to hormones and stress [62]. Thus, five of them contained *cis*-elements sensitive to auxin (ACTTTA, CATATG, TGTCTC) [37], gibberellin (TAACAAA and TTTTTTCC), ABA (AWTTCAAA). The level of PO mRNA in plants increased on treatment with abscisic, jasmonic, and salicylic acids [47, 63]. A positive regulation of anionic POs genes expression under the influence of ABA has been described. But it is also known that high concentrations of ABA exclude the hypersensitivity reaction of cells induced by elicitors, salicylic, and arachidonic acids [64] selectively suppressing activities of anionic POs in wheat [65] and expression of the pathogen-induced PO genes in tobacco and tomato [58]. In promoters of the genes of *Arabidopsis* PO ABRE-elements were found [62].

A *trans*-factor ARF has been identified that is highly specific to an auxin-sensitive region of the promoter (TGTCTC) in the PO gene [28]. Many PO genes are found to have *cis*-elements (YTGTCWC) capable to reacting to water stress, hypothermia, or infection. It is supposed that expression of the zucchini gene encoding the pectin-bound PO is controlled by auxins [66]. Plants hyperproducing IAA are characterized a high PO activity and, as a result, an increased resistance under stress [67]. However, mutants insensitive to endogenous auxin are susceptible to infection with pathogens [68]. Interestingly, reaction of plants to exogenous auxin is just the opposite. It is shown that IAA inhibits PO activity by oxidation a number of phenolic compounds but induces catalysis of its own oxidation [69]. It is worth noting that the PO molecule exposes on its surface a site of auxin reception. Homologous motifs of amino acid sequence on ABP1 protein binding to IAA but differing from amino acid sequence of the active site cleaving hydrogen peroxide are found in PO [50] and in proteins of the germin family [70]. On the surface of the PO molecule the site is located within the distal domain remote from the active site of the enzyme and contains the residue Trp117 [50]. IAA oxidation can occur in different ways and has a free radical character. Co-factors of the process are the Mn^{2+} ions, and a number of phenolic compounds [27, 71]. PO localized in the cell wall play an important role in the oxidation of IAA. According to Hamburg [72], they are described as the main protector barrier, preventing the entry of exogenous auxin in the cell. It should be noted that many plant pathologist associated the aggressiveness of phytopathogenic fungi with an active synthesis of auxin-like compounds. Hence, at least two mechanisms regulating PO activity exist with involvement of auxins related to their effect on the enzyme activity and gene expression. Thus, although some regulatory mechanisms of PO gene expression are known, the interaction of auxins with PO genes and their products requires further studies.

From the latex of the *Euphorbia characias*a Ca^{2+}-dependent PO was isolated that interacted with calmodulin at two binding sites [73], the PO *ELP* gene was sequenced, and positions of two calmodulin-binding sites were determined [74]. The interaction calmodulin with ELP PO was shown both experimentally and using computerized programs. The authors have also shown that the Ca^{2+}/calmodulin ratio regulates (increases) the ELP PO activity and mechanisms of the protective response of the plant.

Salicylic and Jasmonic Acid in Regulation of Peroxidase Activity

Plants constantly interact with pathogens realizing different types of nutrition. Necrotrophic pathogens kill infected cells of the host, whereas biotrophs can live long time in tissues and suppress their immune responses. Hemibiotrophic forms of pathogens can combine strategies of necrotrophs and biotrophs. Depending on the pathogen trophicity plants activate defense mechanisms that are often radically different. The mechanism of their regulation is widely discussed. It includes salicylic (SA) and jasmonic (JA) acids [75], which induce, respectively, systemic acquired resistance (SAR) to biotrophs [76, 77] and systemic induced resistance (ISR) to phytophages (insects) and necrotrophic pathogens [78]. The resistance of potato is developed under the influence of both SA and JA to late blight [79]. In the early 1980's, it was first established that SA is involved in the development of SAR, which effective marker is the expression of gene encoding PR-1 [80]. Some isogenes of class III PO are also sensitive to SA [81]. JA is involved in ISR triggering by activation of genes encoding enzymes of phenolics metabolism, including some isoPO [82] and proteinase inhibitors (for example, PDF1.2 and PR-6). In particular, JA activated two SA-insensitive isogenes of POs in *S. humilis* plants.

It was reported that signals inducing SAR and ISR are strictly individual and often interfere with each other. Therefore, the combine addition of SA and JA is less efficient in the improvement of plant defence than individual compounds [83, 84]. If JA treatment multiple activates POin the leaves of wild-type Arabidopsis, in the *cep1* mutant, with constitutively increased SA production, this is not observed [85]. However, the cases of the absence of SA and JA mutual interference are also known, when these signal compounds inducing SAR and SIR behave additively or even as synergists [86, 87].

A diversity of responses of PO genes involved in SAR and ISR formation is evidently related to a wide spectrum of molecular forms of this enzyme. Physiological functions of its particular isoforms should be identified [82, 88] because PO, especially anionic isozymes of PO, are often referred to as key factors in disease resistance of plants [89, 90].

One of the evidences of the PO regulation during pathogenesis is the inhibition of the oxidative burst induced by SA in plants treated by PO inhibitors – salicylhydroxamic acid or monodehydroascorbate [91]. Proteomic analysis of *Pisum sativum* apical meristem showed that ascorbate PO neutralizing H_2O_2 in ascorbate-glutathione cycle was induced by SA. However, apoptosis-inducing concentration of SA (100 μM) reduced the content of this enzyme [92]. Apparently, for triggering of SA-induced genes the oxidative burst caused by pathogen penetration is necessary. So, in SA-treated asparagus plants, infected by root rot pathogen the increase of PO activity was found, but individual influence of SA or pathogen did not lead to this observation [93].

A variety of class III PO genes are SA-induced, although the majority of them are not SA-competent, in the strict sense. It is expected that sensibility to SA is a specific "marker" of PO, participating in the defence reactions [80]. So, in our investigations in wheat plants under the influence of SA the activity of PO with p*I*~ 3.5 and p*I* ~ 9.7 in cytoplasmic fractions of proteins increased. Interestingly, these isoPO can interact with some surface structures of pathogenic fungi, example, chitin and glucans [94]. Recently we found that SA promoted transcription of genes encoding anionic PO in wheat, infected by pathogenic fungus *Septoria* *nodorum* [95]. The treatment of parsley calli by synthetic analogue of SA (benzothiodiasole) induced the expression of anionic isoPO [96].

It should be noted that SA can stimulate the capacity of PO to produce O_2^- and inhibit the antioxidant activity of that simultaneously [51, 80]. It causes the SA-radicals generation and production of some phenolic compounds participating in antibiotics synthesis (phytoalexines, terpenes, alcaloides) [97, 98] and lignin polymerization. SA-radical (or SA) can take part in lipid peroxidation [97, 99] and SA-induced expression of PR1 gene was suppressed by diethyldithiocarbamate converting peroxy- derivatives of fatty acids in hydroxylic forms [100].

SA-induction of the defense reaction in plants is based on direct mechanism or mediated by secondary messengers, in particular, ROS and products of their action. It was found that SA takes part in the regulation of NADP(H)-oxidase activity and NO level, and can activate lipoxygenase- and MAP-kinase systems [83, 101]. Furthermore, a number of researches propose to classify the defense reaction on SA-dependent and SA-independent pathways, influencing reciprocally on each other and provide an integrated network of regulatory cross-talks [51, 102].

Among wide range of plant POs an anionic isoforms have special interest. So, in tobacco plants superpducing the POs with p*I* ~ 3.0 and ~ 3.5 greatly increases lignification of tissues of the stem and leaves [16]. Participate in the formation of anionic POs suberin was confirmed by iron bean plants do not normally synthesize suberin. [74] Significant activation of anionic PO isozymes was observed in the core wounding of tobacco. Activation of the anionic PO with p*I* ~ 4,4 and its allocation to the extracellular space was found in infected with *Phytophthora capsici* sustainable pepper plants [103]. Increasing of the specific activity of PO isoforms with p*I* ~ 5,6 and ~ 6.1 in highly sensitive varieties of tobacco leaves inoculated with tobacco mosaic virus. However, the wounding of leaves wasn't induced an activity of these isoenzymes [34]. It can be possible that this PO catalyzing the synthesis of lignin near plasmodesmata, and directly restrict the expansion of viral infection on them. [93] Thus, changes in total PO activity in plants infected by viruses are also important.

Interestingly, in plants POs induced SA or JA are presented. Besides, the same genes encoding POs are susceptible in various degrees for both this compounds.

So, it is known that POs belong to a proteins induced by wounding, as well as PR-6 (proteinase inhibitor II). It has been shown that a significant increase in the expression of the latter against the wounding can be caused by exogenous JA and its analogues. Although an expression of the gene encoding wound-induced PO tpoxN1 not significantly increased under the influence of JA, as well as it was independent of the SA-content in plants [58]. In rice exogenous application of JA activated an expression of 6 of the 10 genes encoding the stress-inducible PO (S14493, R2576, R2184, R2693, prxRPN and S11222). Under the influence of SA only 4 of them (S14493, R2184, R2693 and prxRPN).

Next, 10 POX genes that responded to fungal inoculation in the leaves were classified by their responses to treatments with PBZ, JA and/or wounding [104]. Thus, the question about regulation of SA and JA of PO activity requires further investigation.

Peroxidase in Lignification and Suberinization of Cell Walls of Plants and Phytopathogens

In addition to cellulose, vascular plants usually contain the unique biopolymer - lignin, which is specific only for them, and its fraction in some trees can reach 30% of their dry weight. The presence of lignin in moss and algae is still under discussion. Therefore, appearance of lignin is often considered to be associated with appearance of terrestrial plants because only they have a "supporting apparatus" and a vascular system necessary for transporting nutrients and water through the tissues [105]. Lignin is also a component of the secondary cell wall and of the intercellular plate in all cells of terrestrial plants that have virtually terminated their growth, and it is responsible for their rigidity required to support the plant's weight and counteract the negative hydrostatic pressure of lifting water across the xylem [106, 107]. The same biopolymer is responsible for the plant's resistance to pathogens and phytophages because it is not degraded by their hydrolases and, thus, creates a physical barrier in the path of their expansion and inhibits diffusion of toxins produced by them. Tissues with hyper synthesis a pathogen- or phytophage (insect)-induced POs, as well as tissues of transgenic plants with high PO activity become not only unattractive for insects but also toxic for them [108].

The lignin accumulation and deposition is one of the important defense mechanisms of plant interaction with pathogens, realized by increase of mechanical strength of plant cell walls, their protection from hydrolysis and destruction by extracellular pathogen enzymes. The lignin does not destruct by fungi and insect hydrolases, with exception ligno-POs some fungi. The plant tissues, exposing to lignification, become inedible for pathogen and heterotrophs. Probably, the monolignols polymerization in a lignin occurs on free-radical mechanism. There is the question about of enzymes which catalyze this process [109, 110]. The urgency of this question is bind with perspective requests of cellulose manufacture as the lignin is one of undesirable biopolymers and difficultly taken from wood [111]. Judging by numerous literary data, in catalysis of polymerization of phenolic monomers in a lignin, a suberin [112] and cutin [113] are involved the apoplastic anionic PO isoforms. Their accumulation after wounding on an internal wall of the periderm by the method immunofluorescence has found [114]. A suberin and anionic POwere localized in a periderm of a potato tuber on an internal part of a cell wall. This proves its participation in suberin formation [112]. Activity of anionic PO and tissue suberinization are elevated addition in a nutrient medium of ions Fe^{3+}. The lignin is synthesized from L-phenylalanine and L-tyrosine through a stage of formation of cinnamic acids. Their condensation leads to formation of dimers, repeatedly exposing to attack POs, forming the dimeric free radicals. These radicals are involved in condensation in more high-oligomerous lignols. The process comes to the end with formation of the disorder three-dimensional polymer named a lignin. These polymers are secreted from cytoplasm in a cell wall. Here they are exposed to oxidation under the influence of connected PO and H_2O_2 in corresponding phenoxylic radicals which are involved in polymerization. Lignin and suberin polymerization occurs not in a random way. This mechanism is organized. The lignin phenolic rings in polymer are focused mainly in a plane of a cell wall with participation a polysaccharide. Here the residues of ferulic acid serve lignin synthesis [115]. Earlier tried to reveal the presence H_2O_2 at a course of lignification in order to confirm the participation of POs in formation of the free radicals of monolignols. Presence

H_2O_2 in cell walls of a poplar and its absence in cambial cells is shown. The other important approach to definition of the PO role in lignification is connected with obtaining of transgene plants with the changed PO activity. Group of M. Lagrimini have received some transgenic lines of tobacco with considerably raised expression of anionic PO gene, presumably participating in lignification, and on the contrary, it's decreased regulation [43]. There are data, that the part of plants with the high level PO produces more lignins in comparison with the control, however the reduction of lignin quantity in plants with lower PO activity is not observed.

Special interest in connection with role of PO in a plant morphogenetic processes and lignin synthesis represents their communication with processes of plant tissue defense from pathogens and phytophages (wound response). The important characteristic of a lignin is that only few of a lot of number of parasitic microorganisms (for example, fungus destroying wood) can cleave it. Therefore lignifications of plant cell walls serve as a barrier on a way of distribution of an infection. The induced anionic PO or transgenic plants with naturally high enzyme activity become toxic for pathogen or phytophage [116]. The lignification of plant cell walls creates a physical barrier on a way of distribution a hyphae of parasitic fungus and blocks the diffusion of allocated toxins. It is established, that the lignin accumulation at protective plant reactions occurs as a result of its synthesis de novo, induced by products of enzymatic destructions of cell walls by pathogens.The artificial inhibition of lignification can lead to disorder of the immune response that has been shown on an example of wheat infection by brand [117]. First of all, it is possible to note high stability of vessels of a conductive tissue to pathogens.Besides, initiation of morphogenetic processes in culture of the plant cells also leads to enhancement of their stability to pathogenic fungi [118]. The intensive generation of the active oxygen forms and the subsequent lignification of their cell walls with participation of anionic pathogen-induced PO at calluses are found out [119]. Similar reaction at not infected calluses has not observed. Fixation of a cell wall in plants occurs through formation in it of intermolecular communications, formation of various polymers, for example, callose to the subsequent and intensive saturation a lignin-like material [120; 121]. There is an opinion, that the specific protective effect of lignification at fungal pathogenesis is connected with creation of a mechanical barrier on a growth way of pathogens, interfering their distribution, protection of a cell wall from hydrolytic enzymes and a cell from the toxins produced by pathogen, limitation of flux to pathogen of host metabolites and water, and in a plant tissues - toxins [122].As consequence of it is observed packeting of infectious structures of a parasite and their destruction.Lignin biosynthesis, as a rule, is observed in places of introduction of infectious structures of a fungus in plant tissues. The formed condensed formations observed at infection have received the name «papilla».They are not typical neither to a plant, nor a fungi and represent the difficult structural formation between plasmalemma of the host and a cell wall of pathogen. Speed of their formation is important and delay of this process leads to disease development [123].The pathogen focuses, surrounded by papilla's, are thinner and often dendritic. Plants cell wall are saturated by contain the ROS, PO, cellulose, callose, silicon, and phenolic compounds and, accordingly, a product of their direct interaction – the polymer [124; 125].

Numerous literature data indicate that polymerization of phenolic monomers into lignin, suberin, and cutin is catalyzed by apoplastic anionic but not by cationic isoPOs [115]. By immunofluorescence their accumulation was shown upon wounding on the inner wall of the periderm, and this accumulation correlated with the accumulation of lignin in the same zone

[126]. In the periderm of a potato tuber the suberin layer and anionic PO were located on an inner part of the suberinated cell wall, and some authors [127] think this to be a manifestation of its involvement in suberin production. Three genes encoding class III POs are known to be involved in the lignifications of cell walls in the spruce *Piceaabies* [110]. Only two of five isolated POs of cottonwood are involved in lignification [128]. In transgenic tobacco inhibition of expression of the *NtPrx60* gene encoding the protein TP60 significantly decreased the content of lignin and production of vascular tissues [129].

Activities of PO isoenzymes and lignin contents change during cytodifferentiation in cell and tissue cultures. Thus, on starting a culture of medic cells, the PO activity increased due to activation of cationic isoforms but in differentiation stage different isoPOs were presented neutral and acidic isoforms [130]. Similar changes in the activity of anionic PO were also observed in *Prosopsis* plants after wounding [42].

The PO genes of *AtPrx42*, *AtPrx64*, and *AtPrx71* were expressed during the formation of secondary cell wall of conducting elements [131]. The *AtPrx53* gene was expressed similarly in vascular bundles [130]. The *AtPrx66* gene was highly homologous to the PO *ZePrx01* gene of zinnia that was expressed only during the differentiation of tracheal elements, and the corresponding protein catalyzes the *in vitro* formation of lignin [132]. The *AtPrx66* gene is expressed in root vessels, and this confirms the involvement of this gene in the lignification of cell walls of the root. A detailed analysis of signaling regulation of the *AtPrx17* gene has shown that the transcription factors AGL15/18 are crucial regulators of its expression. Another three genes (*AtPrx13*, *AtPrx30*, and *AtPrx55*) were mainly expressed in flowers with participation the *trans* factors SHP1 and SHP2 regulating lignification of the pod dehiscence zone [133].

Peroxidases Can Sorb on Polysaccharides of Phytopathogens Cell Wall

In 1958 Siegel showed that the addition of chitin to the reaction mixture for oxidation and polymerization, with the help of the plant PO of the phenol compound eugenol present inlignin, increased the output of the reaction products and changed their structural quality as compared with the reaction without chitin. Hammerschmidt and Kuc' [134] observed the formation of lignin on the *Colletotrichum lagenarium and C. cucumerinum* fungal mycelium surface in the presence of PO extracted from cucumber plants, eugenol or coniferyl alcohol and hydrogen peroxide. Consequently, on the surface of these pathogens, a hypothetical “anchor” molecule for lignin synthesis initiation should be present, similar to pectin in the plant cell wall. Both plant tissues and pathogen hyphae may be lignified. The mechanisms of targeted lignification in the vicinity of infecting fungal structures, as well as the structures of plant cell walls, are poorly understood.

It has been shown that the cell walls of the majority of fungi contain chitin [135]. Chitin can be extracted in great quantities by industrial processes from crab and shrimp shells or the larvae cuticles of insects such as *Musca domestica* or *Galleria mellonella* [136]. Some authors suggest a scheme of chitin extraction from the fungal mycelium of moulds (*Aspergillus sp., Mucor sp.*) [137]. The low price and ecological safety of this biopolymer define it as an available matrix for technological processes. In this section, we would like to demonstrate the possibility of using chitin for the isolation of certain isoPOs from a great number of plants.

Based on these observations, it was proposed that chitin can be considered as a matrix for plant lignin formation. Using chitin as a biological matrix, we have developed an approach for isolating PO from the protein extracts of plants [23]. Initially, we found that a fraction from wheat extract absorbed on chitin contained a dark-brown pigment which could be eluted with 1 M NaCl and which manifested the enzymatic activity of PO in the presence of H_2O_2. For the first time proteins with p*I*~ 3.5 and a molecular weight of 37 kDa were found in plants, and these proteins could be absorbed by chitin due to ion exchange and manifest PO activity [24]. Next, we detected POs with similar properties in a great number of plant species After this, we demonstrated the ability of wheat anionic POs to bind to the chitin of the cell walls of fungal pathogens. We called these POs "chitin-binding POs" [138]. We were the first to demonstrate the binding of the anionic PO of wheat root to chitin. Besides this, we observed that in some species the activity of POs was increased in the unbound (*Armoracia rusticana, Lagenaria siceraria)* or eluted (*Pisum sativum, Galega orientalis, Brassica oleraceae)* fractions of proteins after interaction with chitin. So, the sorption of PO on polysaccharides was not a classic ion-exchange interaction because the proteins were different in both isoelectric points and the molecular weights exhibited affinity for them. This conclusion was confirmed by the fact that the desorption of PO was facilitated by increasing NaCl concentrations as well as that isoPOs with a different p*I* can bind chitin, as has been shown earlier [139]. Additionally, it was shown that the ion-exchange affinity of POs for polysaccharides is determined by the presence of zones with high electrostatic potential in the enzyme molecule [140]. Therefore, it can be assumed that the wheat POs adsorbed on chitin contain sites that can specifically interact with their acetyl residues. We found that the PO activity in all the plants tested increased in the presence of chitin as compared with the activity in crude extracts. This activation was not only found in chitin-eluted fraction but also in that fraction not adsorbed on chitin. This data confirms the possibility of PO activation in the presence of polysaccharides [141]. This increase of the activity is probably due to the conformational change of the enzyme molecules [142] or the clearing out from the extracts of various inhibitory factors possessing a higher affinity with chitin and remaining bound to it even after elution with 1 M NaCl [139].

Obviously, defense reactions of plants in response to pathogens injection are complicated and multifactorial. An important physiological sense of these reactions is the strengthening of the cell wall and construction on the pathogens' paths of physical barriers consisting of lignin- and suberin-containing polymers, which also possess high toxicity. This defense reaction is developing more or less automatically, thus these barriers are formed just in the zone of pathogen penetration and concurrently with the beginning of the active pathogen expansion into the tissues of the host plant. However, mechanisms of development of these events are still unclear. Possibly, plants contain a whole subclass of isoPOs and their encoding genes. This subclass should be characterized by the ability for binding with polysaccharides, in particular with chitin, and indirectly display the defense functions directed to strengthen the cell wall of the host and to isolate with lignin the non-infected host tissues from the pathogen.

Conclusion

The PO performs vitally important functions in the plant cell mainly associated with oxidation of phenolic compounds and with formation and strengthening of the cell wall. PO is involved in the oxidative transformation of molecules with growth-regulating or signaling activities and, as a result, can also perform regulatory functions in the cell. These physiological functions of the enzyme are especially important in the case of cell damage under exposure to various stress factors, including infection with fungal pathogens. The molecular and functional heterogeneity makes it possible to change the ratio between the isoforms to the advantage of those that are more appropriate to specific stress conditions of the environment, and an increase in the PO activity can be considered as a protective function of the organism. Functions of PO can be associated not only with synthetic processes during cell differentiation and organogenesis, but also with regulation of plant cell metabolism and control of plant growth and development. However, it is still difficult to understand why the same PO isoforms can be responsible for plant growth and for the oxidative burst in leaves during fungal pathogenesis.

The available information points out the need for determination of role of oxidoreductase in plant resistance to unfavorable environmental factors and consideration of their role in the concentration, generation, and utilization of ROS in the infection zone. It is suggested that both quantitative and qualitative changes in the level and activity of oxidoreductases can lead to changes in reactions of free radical oxidation. Therefore, ROS and oxidoreductases involved in the system of their generation and degradation can be in total combined into the pro-/antioxidant system [143].

Genes encoding the class III POs are present in all higher plants, forming a multigenic protein family. But until now the relationship between isoPOs and genes encoding them has not been studied in detail, apparently because of methodical difficulties of such analysis. Nevertheless, such data recently began to appear, and this simplifies the explanation of structural–functional characteristics of PO.

Some POs are functionally associated with the plant cell wall and contribute to its modification due to a high affinity for hemicellulose (most likely, to pectin). Some isoPOs can bind with some components of the cell wall of pathogenic fungi and plants. It is likely that PO binding with polysaccharides performs a signaling and protective function in plants due to immediate involvement in the action of the pro-/antioxidant system and to triggering some signaling systems of the plants [144]. The possibility of regulating the PO-encoding genes expression by different regulators of plant resistance including oligosaccharides allows determining the role of the enzyme in plant immunity.

Obviously, defense reactions of plants in response to pathogens injection are complicated and multifactorial. An important physiological sense of these reactions is the strengthening of the cell wall and construction on the pathogens' paths of physical barriers consisting of lignin- and suberin-containing polymers, which also possess high toxicity. This defense reaction is developing more or less automatically, thus these barriers are formed just in the zone of pathogen penetration and concurrently with the beginning of the active pathogen expansion into the tissues of the host plant. However, mechanisms of development of these events are still unclear. Possibly, plants contain a whole subclass of isoPOs and their encoding genes. This subclass should be characterized by the ability for binding with polysaccharides, in particular with chitin, and indirectly display the defense functions directed to strengthen the cell wall of the host and to isolate with lignin the non-infected host tissues from the pathogen.

Acknowledgments

The authors are grateful for financial support from the Ministry of Education and Science of Russian Federation P339.

References

[1] Almagro, L., Gomez Ros, L. V.,Belchi-Navarro, S., Bru, R., Ros Barcello, A., and Pedreno, M. A. // *J. Exp. Bot.* 2009. V. 60. P. 377-390.
[2] Allen, R. G., and Tresini, M. // *Free Radic. Biol. Med.* 2000. V. 28. P. 463-499.
[3] Heitefuss, R. // *Naturwissenschaften.* 2001. V. 88. P. 273-283.
[4] Tarchevsky, I. A. *Signaling Systems of Plants* [in Russian], Nauka, Moscow. 2002.
[5] Mahalingam, R., and Fedoroff, N. // *Physiol. Plant.* 2003. V. 119. P. 56-68.
[6] El-Sayed, M., and Verpoorte, R. //*Plant Growth Regul.* 2004. V. 44. P. 53-58.
[7] Van Loon, L. C., Rep, M., and Pieterse, C. M. J. // *Ann. Rev. Phytopathol.* 2006. V. 44. P. 135-162.
[8] Ruleva, N. Yu., Zvyagintseva, M. A., and Dugin, S. F. // *Sovrem. Naukoemkie Tekhnol.* 2007. V. 8. P. 1-5.
[9] Morkunas, I., and Gmerek, J. // *J. Plant Physiol.* 2007. V. 164. P. 185-194.
[10] Felle, H. H., Kondorosi, E. A., Kondorosi, A. A., and Schultze, M. // *Planta.* 1999. V. 209. P. 207-212.
[11] Drew, M. C., He, C.-J., and Morgan, P. W. // *Trends Plant Sci.* 2000. V. 5. P. 123-127.
[12] Maksimov, I. V., Cherepanova, E. A., Murtazina, G. F., and Chikida, N. N. *Izv. Ros. Akad. Nauk.* 2006. V. 33. P. 575-580.
[13] Kaothien, P., Shimokawatoko, Y., Kawaoka, A., Yoshida, K., and Shinmyo, A. // *Plant Cell Rep.* 2000. V. 19. P. 558-562.
[14] Caruso, C., Chilosi, G., Caporale, C., Leonardi, L., Bertini, L., Magro, P., and Buonocore, V. // *Plant Sci.* 199.V. 140. P. 87-97.
[15] Ghosh, M. // *Ann. Botany.* 2009. V. 98. P. 1145-1153.
[16] Akhunov, A. A., Golubenko, Z., Khashimova, N. R., Mustakimova, E. Ch., and Vshivkov, S. O. // *Chem. Nat. Comp.* 2008.V. 44. P. 493-496.
[17] Ye, X. Y., and Ng, T. B. // *Life Sci.* 2002. V. 71. P. 1667-1680.
[18] Yang, Y., and Anderson, E. J. // *J. Appl. Microbiol.* 2006. V. 86. P. 211-220.
[19] Sarowar, S., Kim, E. N., Kim, Y. J., Ok, S. H., Kim, K. D., Hwang, B. K., and Shin, J. S. // *Plant Sci.* 2005. V. 169. P. 55-63.
[20] Minibaeva, F. V., and Gordon, L. Kh. // *Rus. J. Plant Phys.* 2003. V. 50. P. 459-464.
[21] Sitbon, F., Hennion, S., Little, C. H. A., and Sundberg, B. // *Plant Sci.* 1999. V. 141. P. 165-173.

[22] Okazaki, Y., Isobe, T., Iwata, Y., Matsukawa, T., Matsuda, F., Miyagawa, H., Ishihara, A., Nishioka, T., and Iwamura, H. // *Plant J.* 2004. V. 39. P. 560-565.
[23] Khairullin, R. M., Yusupova, Z. R., and Maksimov, I. V. // *Fiziol. Rast.* 2000. V. 47. P. 108-113.
[24] Maksimov, I. V., Cherepanova, E. A., Yarullina, L. G., and Akhmetova, I. E. // *Appl. Biokhim. Mikrobiol.* 2005. V. 41. P. 616-620.
[25] Maksimov, I. V., Cherepanova, E. A., and Surina, O. B. // *Rus. J. Plant Phys.* 2010. V. 57. P. 131-138.
[26] Herder, I., Lievens, S., Rombants, S., Holsters, M., and Goormachting, S. // *Plant Physiol.* 2007. V. 144. P. 717-727.
[27] Nanda, A. K., Andrio, E., Marino, D., Paulu, N., and Dunand, C. // *J. Integr. PlantBiol.* 2010. V. 52. P. 195-204.
[28] Glyanko, A. K., and Vasil'eva, G. G. // *Appl. Biokhim. Mikrobiol.* 2010. V. 46. P. 21-38.
[29] Coego, A., Ramirez, V., Ellul, P., Mayda, E., and Vera, P. // *Plant J.* 2005. V. 42. P. 283-293.
[30] Choi, H. W., Kim, Y. J., Lee, S. C., Hong, J. K., and Hwang, B. K. // *Plant Physiol.* 2007. V. 145. P. 890-904.
[31] Hilaire, E., Young, S. A., Willard, L. H., McGee, J. D., Sweat, T., Chittoor, J. M., Guikema, J. A., and Leach, J. E. // *Mol. Plant–Microbe Interact.* 2001. V. 14. P. 1411-1419.
[32] Chassot, C., Nawrath, C., and Metraux, J. P. // *Plant J.* 2007. V. 49. P. 972-980.
[33] Little, D., Gouhier-Darimont, C., Bruessow, F., and Reymond, P. // *Plant Physiol.* 2007. V. 143. P. 784-800.
[34] Mohr, P. G., and Cahill, D. M. // *Funct. Integrat. Genomics.* 2007. V. 7. P. 181-191.
[35] Cheong, Y. H., Chang, H. S., Gupta, R., Wang, X., Zhu, T., and Luan, S. // *Plant Physiol.*2002. V. 129. P. 661-677.
[36] Cai, S., and Lashbrook, C. C. // *Plant Physiol.* 2008. V. 146. P. 1305-1321.
[37] Cosio, C., and Dunand, C. // *J. Exp. Botany.* 2009. V. 60. P. 391-408.
[38] Hammond, J. P., Bennett, M. J., Bowen, H. C., Broadley, M. R., Eastwood, D. C., May, S. T., Rahn, C., Swarup, R., Woolaway, K. E., and White, P. J. // *Plant Physiol.* 2003. V.132. P. 578-596.
[39] Klok, E. J., Wilson, I. W., Wilson, D., Chapman, S. C., Ewing, R. M., Somerville, S. C., Peacock, W. J., Dolferus, R., and Dennis, E. S. // *Plant Cell.* 2002. V. 14. P. 2481-2494.
[40] Yokoyama, R., and Nishitani, K. // *J. Plant Res.* 2006. V. 119. P. 189-194.
[41] Rouet, M. A., Mathieu, Y., and Lauriere, C. // *J. Exp. Botany. 2006.* V. 57. P. 1323-1332.
[42] Lehner, G., and Cardemil, L. // *Tree Physiol.* 2003. V. 23. P. 443-452.
[43] Dowd, P. F., and Lagrimini, L. M. // *Transgen. Res. 2006.* V. 15. P. 197-204.
[44] Bolwell, G. P., Blee, K. A., Butt, V. S., Davies, D. R., Gardner, S. L., Gerrish, C., Minibayeva, F., Rowntree, E. G., and Wojtaszek, P. // *Free Radic. Res.* 1999. V. 31. P. 131-145.
[45] Liszkay, A., Kenk, B., and Schopfer, P. // *Planta.* 2003. V. 217. P. 658-667.
[46] Minibayeva, F., Mika, A., and Luthje, S. // *Protoplasma.* 2003. V. 221. P. 67-72.

[47] Crockard, A., Bjourson, A., and Cooper, J. // *Mol. Plant–Microbe Interact.* 1999. V. 12. P. 825-828.

[48] Repka, V. // *ActaVirol.* 2000. V. 44. P. 249-257.

[49] Chen, C., Baucher, M., Christensen, J. H., and Boerjan, W. // *Euphytica.* 2001. V. 118. P. 185-195.

[50] Savitsky, P. A., Gazaryan, I. G., Tishkov, V. I., Lagrimini, L. M., Ruzgas, T., and Gorton, L. // *Biochem. J.* 1999. V. 340. P.579-583.

[51] Kawano, T., and Furuichi, T. in *Salicylic Acid: A Plant Hormone* (Hayat, S., and Ahmad, A., eds.) Springer, Berlin-Heidelberg. 2007. P. 277-322.

[52] Jusupova, Z. R., Mullagaliev, I. R., and Maksimov, I. V. *Abst. 2nd Int. Symp. Plant Growth Substances: Intracellular Hormonal Signaling and Applying in Agriculture.*Kyiv, Ukraine. 2007. P. 158.

[53] Asselbergh, B., Achuo, A. E., Hofte, M., and van Gijsegem, F. *Mol. Plant Pathol. 2008. V.*9. P. 11-24.

[54] Bellaire, B. A., Carmody, J., Braud, J., Gossett, D. R., Banks, S. W., Lucas, M. C., and Fowler, T. E. // *Free Radic. Res.* 2000. V. 33. P. 531-545.

[55] Sasaki, K., Yuichi, O., Hiraga, S., Goton, Y., Seo, S., Mitsuhara, I., Ito, H., Matsu, H., and Ohashi, Y. *Mol. Genet. Genomics.* 2007. V. 278. P. 709-722.

[56] Lavania, M., Chauhan, P. S., Chauhan, S. V. S., Singh, H. B., and Nautiyal, C. S. // *Cur. Microbiol.* 2006. V. 52. P. 363-368.

[57] Zamocky, M. // *Eur. J. Biochem.*2004. V. 271. P. 3297-3309.

[58] Hiraga, S., Ito, H., Yamakawa, H., Ohtsubo, N., Seo, S., Mitsuhara, I., Matsui, H., Honma, M., and Ohashi, Y. //*Mol. Plant–Microbe Interact.* 2000. V. 13. P.210-216.

[59] Passardi, F., Bakalovic, N., Teixeira, F. K., Pinheiro-Margis, M., Penel, C., and Dunand, C //. *Genomics.* 2007. V. 89. P. 567-579.

[60] Mlickova, K., Luhova, L., Lebeda, A., Mieslerova, B., and Pec, P. // *Plant Physiol. Biochem.* 2004. V. 42. P. 753-761.

[61] Yannarelli, G. G., Gallego, S. M., and Tomaro, M. // *Envir. Exp. Botany.* 2006. V. 56. P. 174-181.

[62] Lebedeva, O. V., Ezhova, T. A., and Shestakov, S. V. //*Dokl. Ros. Akad. Nauk.* 2008. V. 394. P. 136-138.

[63] Ueeda, M., Kubota, M., and Nishi, K. // *Physiol. Mol. Plant Pathol.* 2006. V. 67. P. 149-154.

[64] Audenaert, K., Meyer, G. B. D., and Hofte, M. M. // *Plant Physiol.* 2002. V. 128. P. 491-501.

[65] Agrawal, G. K., Rakwal, R., and Jwa, N.-S. // *Plant Sci.* 2002. V. 162. P. 49-58.

[66] Dunand, C., de Meyer, M., Crevecoeur, M., and Penel, C. // *Plant Physiol. Biochem.* 2003. V. 41. P. 805-811.

[67] Sitbon, F., Hennion, S., Little, C. H. A., and Sundberg, B. // *Plant Sci.* 1999. V. 141. P. 165-173.

[68] Napier, R. // *Ann. Botany.* 2004. V. 93. P. 227-233.

[69] Rogozhin, V. V. *Peroxidase as a Component of the Antioxidant System of Living Organisms* [in Russian]. GIORD. St. Petersburg. 2004.

[70] Mayda, E., Marques, C., Conejero, V., and Vera, P. // *Mol. Plant–Microbe Interact.* 2000. V. 13. P. 23-31.

[71] Mohr, P. G., and Cahill, D. M. // *Funct. Integrat. Genomics.* 2007. V. 7. P. 181-191.

[72] Zamocky, M. // *Eur. J. Biochem.*2004. V. 271. P.3297-3309.

[73] Medda, R., Padiglia, A., Longu, S., Bellelli, A., Arcovito, A., Cavallo, S., Pedersen, J. Z., and Floris, G. // *Biochemistry*. 2003. V. 42. P. 8909-8918.

[74] Mura, A., Medda, R., Longu, S., Floris, G., Rinaldi, A. C., and Padiglia, A. // *Biochemistry*. 2005. V. 44. P. 14120-14130.

[75] Glazebrook, J. // *Ann. Rev. of Phytopathology*. 2005. V. 43. P. 205–227.

[76] Park, S.-W., Kaimoyo, E., Kumar, D., Mosher, S., Klessig, D.F. // *Science*. 2007. V. 318. N. 5847. P. 113-116.

[77] Halim, V.A., Eschen-Lippold, L., Altmann, S., Birschwilks, M., Scheel, D., Rosahl, S. // *Mol. Plant-Microbe Interact.* 2007. V. 20. P. 1346-1352.

[78] Makandar, R., Nalam, V., Chaturvedi, R., Jeannotte, R., Sparks, A.A., Shah, J. // *Mol. Plant-Microbe Interact*. 2010. V. 23. N. 7. P. 861–870.

[79] Thaler, J. S., Humphrey, P.T., Whiteman, N.K.// *Trends in Plant Science*. 2012. V. 17, No. 5. P. 260-270.6.

[80] Almagro, L., Gomes Ros, L.V., Belchi-Navarro, S., Bru, R., Ros Barcelo, A., Pedreno, M.A. // *J. Exp. Botany*. 2009. V. 60. N. 2. P. 377-390.

[81] Cosio, C., Dunand, C. // *J. Exp. Botany*. 2009. V. 60. P. 391-408.

[82] Curtis, M. D., Rae, A. L., Rusu, A. G., Harrison, S. J., Manners, J. M. // *Mol. Plant-Microbe Interact*. 1997. V. 10. P. 326-338.

[83] Robert-Seilaniantz, A., Grant, M., Jones, J.D. // *Annu. Rev. Phytopathol*. 2011. V. 49. P. 317-343.

[84] Smith, J.L., De Moraes, C.M., Mescher, M.C. // *Pest. Manag. Sci.* 2009. V. 65. P. 497-503.

[85] Traw, M. B., Kim, J., Enright, S., Cippolini, D. F., Bergelson, J. // *Mol. Ecology*. 2003. V. 12. 1125–1135.

[86] Mur, L.A.J., Kenton, P., Atzorn, R., Mierch, O., Wasternack, C. // *Plant Physiology*. 2006. V. 140. P. 249 - 262.

[87] Wees, S., Swart, E., Pelt, J. A., van Loon, L. C., Pieterse, C. M. J. // *Proc. Nat. Acad. Sci.* 2000. V. 97(15). P. 8711– 8716.

[88] Mathé, C, Barre, A, Jourda, C, Dunand, C. // *Arch. Biochem. Biophys*. 2010. V. 500(1). P. 58-65.

[89] Dowd, P.F., Johnson, E. T., Pinkerton, T.S. // *Phytochemistry*. 2010. V. 71. P. 1289–1297.

[90] Maksimov, I. V., Sorokan', A. V., Chereoanova, E. A., Surina, O. B., Troshina, N. B., Yarullina, L. G. // *Rus. J. Plant Phys.* V. 58. P. 299 – 306.

[91] Kawano, T., Muto, S. // *J. Exp. botany*. - 2000. V. 51(345). P. 685 - 693.

[92] Tarchevsky, I.A., Yakovleva, V.G., Egorova, A.M. // *Biochem. (Mosc)*. 2010. V. 75(5). P. 590-597.

[93] He, C. Y., Wolyn, D. J. // *Plant Pathol*. 2005. V. 54. P. 227–232.

[94] Maksimov, I. V., Cherepanova, E. A., Yarullina, L. G., Akhmetova, I. E. // *Appled Biochem. Mikrobiol*. 2010. V. 41. P. 616-620.

[95] Burhanova, G.F., Yarullina, L. G., Maksimov, I. V. // *Rus. J. plant physiol*. 2007. V. 54. P. 104 -110.

[96] Katz, V.A., Thulke, O.U., Conrath, U. // *Plant Physiol*. 1998. V. 117(4) P. 1333-1339.

[97] Okazaki, Y., Isobe, T., Iwata, Y. // *Plant J*. 2004. V. 39. P. 560-565.

[98] Reszka, K.J., Britigan, L.H., Britigan, B.E. // *J. Pharmacology and Exp. therapeutic.*2005. V. 315. P. 283–290.

[99] Hatamzadeh, A., Hatami, M., Ghasemnezhad, M. // *African J. Agricultural Res.* 2012. V. 7. P. 540-545.

[100] Kawano, T. // *Plant Cell Reports.*2003. V. 21. P. 829–837.

[101] Karuppanapandian, T., Moon, J.-C., Kim, C., Manoharan, K., Kim, W. // *Australian J. Crop Science*. 2011. V. 5(6). P. 709-725.

[102] Halim, V.A., Altmann, S., Ellinger, D., Eschen-Lippold, L., Miersch, O., Scheel, D., Rosahl, S. // *Plant J.* 2009. V. 57(2). P. 230-242.

[103] Forchetti, G., Masciarelli, O., Alemano, S., Alvares, D., Abdala, G. //*Appl. Microbiol. Biotechnol*. 2007. V. 76. P. 1145 – 1152.

[104] Katsutomo, S., Takayoshi, I., Susumu, H., Katsushi, K., Shigemi, S., Ichiro, M., Atsushi, M., Masataka, I., Hiroyuki, I., Hirokazu, M. and Yuko, O. // *Plant and Cell Physiol.* 2005. V. 45. P. 1442-1452.

[105] Brundrett, M. C. // *New Phytol.*2002. V. 154. P. 275-304.

[106] Sharova, E. I. // *Cell Wall of Plants* [in Russian], St. Petersbug University Publishers, St. Petersburg. 2004.

[107] Karkonen, A., and Kautaniemi, S. // *J. Integr. Plant Biology*. 2010. V. 52. P. 176-185.

[108] Behle, R. W., Dowd, P. F., Tamez-Guerra, P., and Lagrimini, L. M. // *J. Econ. Entomol.* 2004. V. 95. P. 81-88.

[109] Hatfield, R., and Vermeris, W. // *Plant Physiol.* 2001. V. 126. P. 1351-1357.

[110] Marjamaa, K., Kukkola, E.M., and Fagerstedt, K.V. // *J. of Exp. Botany*. 2009. V. 60. P. 367-376.

[111] Chen, C., Baucher, M., Christensen, J.H., and Boerjan, W. // *Euphytica.* 2001. V. 118. P. 185–195.

[112] Bernards, M.A., Fleming, W.D., Llewellyn, D.B., Priefer, R., Yang, X., Sabatino, A., Plourde, G.L. // *Plant physiol.* 1999. V.121. P.135-145.

[113] Carpin, S., Crevecoeur, M., Greppin, H., and Penel, C. // *Plant Physiol.* 1999. *V.* 120. P. 799–810.

[114] Quiroga, M., Guerrero, C., Botella, M.A., Barcelo, A., Amaya, I., Medina, M.I., Alonso, F.J., Forchetti, S., Tigier, H., and Valpuesta, V. // *Plant Physiol.* 2000. V. 122. P. 1119–1127.

[115] Ralph, J., Bunzel, M., Marita, J.M., Hatfield, R.D., Lu, F., Kim, H., Schatz, P.F., Grabber, J.H., Steinhart, H. // *Phytochem. Rev*. 2004. V. 3. P. 79–96.

[116] Ralph, J., Bunzel, M., Marita, J.M., Hatfield, R.D., Lu, F., Kim, H., Schatz, P.F., Grabber, J.H. and Steinhart, H. // *Phytochem. Rev*. 2007. V. 3. P. 79–96.

[117] Moldenhauer, J., Moerschbacher,B.M., and van der Westhuizen, A.J. // *Plant Pathology*. 2006. V. 55. P. 469–474.

[118] Troshina, N.B., Maksimov, I.V., Surina, O.B., and Khairullin, R.M. // *Biol. Bullet.* 2000. V. 3. P. 377-381.

[119] Maksimov, I.V, Troshina, N.B., Yarullina, L.G., Surina, O.B., Jusupova, Z.R., and Cherepanova, E.A. // In “Biotechnology and agriculture and the food industry” ed. G.E. Zaikov. NewYork: Nova Science Publ. 2004. P. 145-150.

[120] Verma, D.P.S., and Hong, Z. // *Plant Mol. Biol.* 2001. V. 47. P. 693–701.

[121] Hawkins, S., and Boudet, A. // *Forest Path.* 2003. V.33. P. 91–104.

[122] Cano-Delgado, A., Penfield, S., and Smith, C. // *Plant J.* 2003. V. 34. P.351-362.

[123] Heitefuss, R. // *Naturwissenschaften.* 2001. V. 88. P. 273–283.
[124] Shinogi, T., Suzuki, T., and Kurihara, T. // *J. Gen. Plant Pathol.* 2003. V. 69. P.7–16.
[125] Humphrey, T.V., Boneta, D.T., and Goring, D.R. // *New Phytologist.* 2007. V.176. P. 7-21.
[126] Quiroga, M., Guerrero, C., Botella, M. A., Barcelo, A., Amaya, I., Medina, M. I., Alonso, F. J., Forchetti, S., Tigier, H., and Valpuesta, V. *Plant Physiol.* 2000. V. 122. P. 1119-1127.
[127] Penel, C., and Dunand, C. in *Signaling in Plants. Signaling and Communication in Plants* (Baluska, F., and Mancuso, S., eds.) Springer-Verlag, Berlin, Heidelberg. 2007. P. 155-171.
[128] Christensen, J. H., Overney, S., Rohde, A., Simon, P., van Motagu, M., and Boerjan, W. // *Plant Mol. Biol.* 2007. V. 47. P. 581-593.
[129] Blee, K. A., Choi, J. W., O'Connell, A. P., Schuch, W., Lewis, N. G., and Bolwell, G. P. // *Phytochemistry.* 2003. V. 64. P. 163-176.
[130] Gazaryan, I. G., Khushpul'yan, D. M., and Tishkov, V. I. // *Usp. Biol. Khim.* 2006. V. 46. P. 303-322.
[131] Yokoyama, R., and Nishitani, K. // *J. Plant Res.* 2006. V. 119. P. 189-194.
[132] Sato, Y., Demura, T., Yamawaki, K., Inoue, Y., Sato, S., Sugiyama, M., and Fukuda, H. // *Plant Cell Physiol.* 2006. V. 47. P. 493-503.
[133] Roeder, A. H. K., and Yanofsky, M. F. // in *The Arabidopsis Book* (Somerville, C. R., and Myerowitz, E. M., eds.) American Societyof Plant Biologists, Rockville. 2006.V. 4. P. 1-50.
[134] Hammerschmidt, R., and Kuc, J. // *Phisiol. Plant. Pathol.* 1982. V. 20. P. 61 – 71.
[135] Bowman, S.M., and Free, S.J. // *Bioessays.* 2006. V. 28(8). P. 799-808.
[136] Ostanina, E.S., Lopatin, C.A., and Varlamov, B.P. // *Biotechnology.* 2007. V. 3. P. 38 – 45.
[137] Wu, T., Zivanovic, S., Draugnon, F. A., Conway, W.S., and Sams C.E. // *J. Agric. Food Chem.* 2005. V. 53. P. 3888-3894.
[138] Maksimov, I. V.; Cherepanova, E. A., and Khairullin, R. M. // *Biochemistry (Mosc).*V. 68(1). P. 111 - 115.
[139] Maksimov, I. V.; Cherepanova, E. A.,Yarullina, L. G., and Akhmetova, I. E. // *Applied Biochem. And Mikrobiol.* 2010. V. 41. P. 616-620.
[140] Dunand, C., Tognolli, M., and Overney, S. // *J. Plant Physiol.* 2002. V. 159. P. 1165-1171.
[141] Gulsen, O., Shearman, R. C., Heng-Moss, T. M., Mutlu, N., Lee, D. J., and Sarath, G. // *Crop Science.* 2007. V. 47. P. 767-774.
[142] Liu, G. S., Sheng, X. Y., Greenshields, D. L., Ogieglo, A., Kaminskyj, S., Selvaraj, G., and Wei, Y. D. // *Mol. Plant-Microbe Interact.* 2005. V. 18. P. 730–741.
[143] Maksimov, I. V., and Cherepanova, E. A. // *Usp. Sovr. Biol.* 2006. V. 126. P. 250-261.
[144] Maksimov, I. V.; Sorokan', A. V.; Cherepanova, E. A.; Surina, O. B.; Troshina, N. B. and Yarullina, L. G. // *Rus. J. of Plant Physiol.* 2010. V. 58. N. 2. P. 299–306.

In: Peroxidases
Editors: Lucas Bogaert and Noah Coppens

ISBN: 978-1-62808-261-6

Chapter III

Phospholipids and Peroxidases in Mitochondrial Apoptosis

Ekaterina A. Korobkova*
Department of Sciences, John Jay College of Criminal Justice, (CUNY), New York, NY, US

Abstract

Phospholipids constitute an essential part of cellular membranes, in particular, inner and outer membranes of mitochondria. They influence the function of mitochondrial proteins and cytosolic proteins associates with mitochondria. Phospholipids regulate the assembly and maintenance of the proteins embedded in the membranes, they modulate enzymes' ox-red properties, electron-transfer properties, kinase activity, and peroxidase activity. Phospholipids thus play a crucial role in all cellular processes such as protein biogenesis, protein import into mitochondria, energy production, and apoptosis. Over the last decade, a significant amount of research efforts has been devoted to mitochondrial apoptosis, in particular, early stages of apoptosis that precede mitochondrial permeability transition pore (PTP) opening and the release of proapoptotic factors into the cytoplasm, specifically cytochrome c (cyt c). Cyt c is a soluble protein located in the intermembrane space of mitochondria whose primary function is the electron shuttle between complexes III and IV of the respiratory chain. Cyt c is attached to the inner mitochondrial membrane (IMM) through a phospholipid molecule, cardiolipin (CL). It was demonstrated that cyt c, whose peroxidase function is negligible in a free state, exhibits a significant peroxidase activity upon binding to CL. This cyt c-CL complex catalyzes the peroxidation of CL. Peroxidized CL stimulates the formation of the permeability pore in the membranes and the release of cyt c into the cytoplasm through these pores. The increase of the peroxidase activity of cyt c in the CL-bound state is thus considered to be a prerequisite for apoptosis. Cells have also developed anti-apoptotic mechanisms. One of these anti-apoptotic functions is executed by a mitochondrial phospholipid hydroperoxide glutathione peroxidase or GPx4, a protein that defends cells from oxidative injury, attenuates cardiac disfunction, and participates in sperm development. GPx4 plays an

* Corresponding author: ekorobkova@jjay.cuny.edu.

essential role in preventing cyt c release from mitochondria. Specifically, it reduces oxidized phospholipids precluding cyt c detachment from CL and consequently suppressing apoptosis. We will describe the structure of cyt c and GPx4 as well as the catalytic cycles of the two peroxidases. We will highlight the recent advances and unresolved issues regarding biochemical mechanisms underlying cyt c-CL interactions. We will discuss the structural characteristics of cyt c-CL interactions responsible for the tight binding, unfolding of cyt c, and the enhancement of its peroxidase activity. We will also discuss how peroxidized CL triggers PTP opening and how the balance between the two peroxidases, CL-bound cyt c and GPx4, affects the early stages of mitochondrial apoptosis.

Introduction

Apoptosis is a programmed cell's death as a part of its regular development. During apoptosis cells fragment into smaller particles that are removed from an organism by phagocytosis. Apoptosis and numerous biochemical pathways associated with it are directly related to the initiation and development of various diseases. The design of many therapeutic agents is based on their interference with the apoptotic networks. Two types of apoptosis have been identified: the extrinsic apoptosis, which is triggered through the activation of the receptors located on the cell surface, and intrinsic or mitochondrial apoptosis. There is also evidence indicating the existence of the inter-connective pathways between the extrinsic and intrinsic apoptotic mechanisms (Li et al., 1998).

Over the last decade, a number of works have demonstrated the involvement of mitochondrial apoptosis in the processes associated with the development of cancer (Fulda et al., 2010). Mitochondrial apoptosis is a very complicated process, which is regulated by a network of mitochondrial proteins. It is believed that the intrinsic apoptosis is controlled largely by B-cell/lymphoma-2 (Bcl-2) family of proteins. This family includes a group of anti-apoptotic proteins and proapoptotic proteins (Youle and Strasser, 2008). Bcl-2 family proteins regulate the release of proapoptotic factors such as cytochrome c, Smac/Diablo, and apoptosis-inducing factor (AIF) from the intermembrane space to the cellular cytoplasm, where they activate caspases, ultimately leading to the cell death.

The mechanism of the release of proapoptotic factors from mitochondria is a subject of debates. Under certain conditions such as Ca^{2+} overload or red-ox stress, a large non-specific pore may form in the IMM, which results in the diffusion of small molecules (< 1.5 kDa) from the extramitochondrial space to the matrix and vice versa (Leung and Halestrap, 2008). Since proteins cannot cross the membrane, they will remain in the matrix and exert a colloidal osmotic pressure, which results in the swelling of the matrix. The unfolding of the cristae, the structural components of the IMM, permits the swelling to occur without the disruption of the membrane. As a result of the matrix swelling and the enlargement of the IMM, the outer mitochondrial membrane (OMM) tears, and the proapoptotic factors leave mitochondria (Kroemer and Reed, 2000). The nature of this permeability transition pore (PTP) is a subject of intensive research. However, the explanations of PTP opening and its structure remain controversial. In a classical model, PTP consists of cyclophilin D (CypD), adenine nucleotide translocator (ANT), voltage-dependent anion channel (VDAC), and Bax protein (Zoratti et al., 2005). CypD is attached to the IMM and is located on the matrix side of mitochondria. ANT is embedded in the IMM and exposed to the matrix side. VDAC also called porin forms pores in the OMM (Mannella, 1992). Bax is a proapoptotic protein of the Bcl-2 family, and it is incorporated into the OMM. Classical PTP opening is inhibited by cyclosporine A (CSA). A number of non-classical mitochondrial membrane channels have been discovered that cannot be inhibited with CSA and whose opening is not stimulated with calcium cations. A number of other ion-selective and non-selective IMM channels have been discovered whose activation is linked to the early stages of apoptosis. Zoratti and colleagues provided an extended description of the major channels located in the IMM, their functions, properties, role in mitochondrial physiology, and relevance to apoptosis (Zoratti et al., 2009).

There is evidence that PTP opening is predominantly involved in necrosis and ischemia/reperfusion (Nakagawa et al., 2005; Baines et al., 2005). In addition, studies show that cyt c release may be observed without the depolarization of the IMM and without the rupture the OMM. These findings point at the existence of a more selective mechanism for the release of apoptotic activators to the cytoplasm than the OMM rupture. Thus, Dejean and coworkers showed that proapoptotic proteins Bax and Bak oligomerize and form so called mitochondrial apoptosis-induced channels (MAC) in the OMM. The diameter of the channel pore is 5 nm, which is sufficient to allow the passage of cyt c through the channel and its release into the cytoplasm. (Dejean et al., 2005; Dejean et al., 2006). Interestingly the authors found that the activity of MAC was completely independent of the component of the classical PTP, ANT and VDAC, as they did not coimmunoprecipitate with Bax in lysates from apoptotic cells. Additionally, there was no inhibiting effect of CSA on the MAC channel, suggesting a mechanism independent of the IMM depolarization. Even though there is a lot of other data consistent with the Bax/Bak model of MAC, the nature of the pore remains obscure, as the direct demonstration of pores made of Bax and Bak proteins in the OMM in cells is technically challenging. Further investigation is required to explain the mechanisms underlying the regulation of the onset of apoptosis by Bcl-2 family proteins.

The most important apoptotic factor is believed to be cytochrome c (cyt c), a soluble protein located in the intermembrane space and attached to the IMM through the interactions with cardiolipin (CL), a constituent of the membrane (Gohil and Greenberg, 2009; Osman et al., 2011). When released into the cytoplasm, cyt c binds Apaf-1 protein to form apoptosome, which in turn activates caspase-9 leading to the destruction of the cell.

Studies show that apoptotic cell death is also associated with mitochondrial fission (Suen et al., 2008). An OPA1 protein located in the IMM is known to play an essential role in mitochondrial morphology. Oligomers of OPA1 hold together cristae junctions of the IMM and thus retain cyt c in mitochondria. It was suggested that cristae remodeling is required for the detachment and release of cyt c. However, electron microscopy studies showed that cristae remodeling occurs after cyt c is released into the cytoplasm (Sun et al., 2007). These findings point at the cyt c-CL binding as a crucial component of the mechanism of retaining cyt c on the IMM and, therefore, regulating apoptosis.

Recent findings have shed more light on the mechanisms of the interactions of cyt c and CL, revealing the increase of the peroxidase activity of the cyt c-CL complex compared with that of cyt c in a free state (Kagan et al., 2005). The complex catalyzes the peroxidation of CL. The oxidized CL disturbs the interactions of cyt c with non-oxidized CL, causing its detachment from the membrane and the ultimate release into the cytoplasm. The increase of the peroxidase activity of cyt c upon its binding to CL is thus considered to be a prerequisite for apoptosis.

Cells have also developed protection mechanisms against reactive oxygen species (ROS) and apoptosis. The defensive mechanisms against ROS involve primarily antioxidant enzymes such as superoxide dismutase, catalase, glutathione reductase, glutathione S-transferase, and glutathione peroxidase (GP). Eight different isoforms of GP have been identified in humans. The only isoform, which functions in mitochondria and reduces lipid-hydroperoxides inside membranes, is phospholipid hydroperoxide glutathione peroxidase or GPx4. Studies show that GPx4 inhibits cyt c detachment from the IMM by reducing the peroxidation level on the phospholipids of mitochondrial membranes. The peroxidase activity of GPx4 thus plays on opposite role to that of cyt c, whose peroxidase activity stimulates the initiation of apoptosis.

It has long been realized that biomembranes not only provide protection for the interior of cells and small cellular compartments, but also play an essential role in the majority of biochemical processes associated with the life cycle of cells. A well-known example of a vital role of membranes is the process of oxidative phosphorylation. In the course of this process, electrons are transported along the chain of proteins embedded in the IMM. As a result of this electron transfer, the ATP molecules are synthesized and the intermembrane potential is produced, which maintains the integrity of the mitochondrial matrix and, as a result, the integrity of the whole mitochondrion.

The major constituents of mitochondrial membranes are phospholipids such as phosphatidylcholine (PC), phosphatidylethanolamine (PE), phosphatidylinositol (PI), phosphatidylserine (PS), phosphatidic acid (PA), phosphatidylglycerol (PG), and CL. CL is a dimeric phospholipid, it is the most important phospholipid of mitochondrial membranes. Its membrane content is 20% of the total lipid composition. CL plays multiple roles in mitochondria. CL stabilizes protein complexes comprising respiratory chain and maintains their enzymatic functions. Complex IV of the oxidative phosphorylation system associates with two CL molecules, complex V binds five CL molecules (Eble et al., 1990), while complex III requires CL to maintain its quaternary structure necessary for proper functioning (Gomez and Robinson, 1999). CL also participates in the protein import into mitochondria, ceramide synthesis, aging, and in the early stages of mitochondrial apoptosis.

A role of CL in mitochondrial apoptosis is very complex as it interacts with and regulates the activity of many proteins, which are directly involved in the control of the permeability of membranes and the release of proapoptotic factors into the cytoplasm. This chapter focuses on the interactions of phospholipids with mitochondrial peroxidases, specifically hemoprotein cyt c and selenoprotein GPx4. We will discuss how the dynamics and the mechanisms of the interactions of these peroxidases with phospholipids affect the onset of mitochondrial apoptosis.

The Function of Cytochrome C (Cyt c) As a Stimulator of Apoptosis

1. The Structure of Cyt c

Cyt c is a small protein, 14,000 Da, located in the intermembrane space of mitohcondria and attached to the IMM. The sequence of cyt c has been solved 50 years ago (Margoliash et al., 1961). It is a 104 amino acid residue chain with a heme covalently held by two cysteine residues at the positions 14 and 17. The 3D structure of cyt c was one of the first protein structures that were solved by an X-ray diffraction technique (Dickerson et al., 1971). Dickerson et al. used X-ray method to a resolution of 2.8 Å to solve the structure of horse heart cyt c and to compare it with the bonito cyt c protein. The polypeptide sequence of 104 residues is enfolded around the heme in two parts. The chain of amino acids 1 to 47 is located on the right side and the chain of amino acids 48 to 91 if positioned on the left side of the heme. Thus the heme stays in a crevice formed by the polypeptide chain such that its edge is exposed to solvent. A chain of residues 92 to 104 form an α-helix that climbs over the top tail of the molecule and back through the right side. Met 80 spreads to the heme from the left side of the polypeptide chain and His 18 ligates to the heme from the right side. Cysteines 14 and 17 stretch from the right side of the chain and form covalent bonds with the heme vinyl groups. Later Fisher and coworkers found that the presence of the porphirin ring attached to cysteine residues is crucial for the stabilization of cyt c native conformation (Fisher et al., 1973). The stability of the heme coordination by Met 80 and His 18 were also intensively investigated in later years. Various techniques have been utilized in order to characterize the strength of the ligands-heme binding and its roles in protein folding kinetics and enzymatic reactions. These techniques include magnetic circular dichroism, electron paramagnetic resonance (Simpkin et al., 1989), fluorescence, UV/Vis, stopped flow technique, and ^{1}H NMR (Colon et al., 1997; Chen et al., 2007). The binding of Met 80 and His 18 to the heme of cyt c is very strong. His 18 (5th ligand) is bound covalently, so it does not detach from the heme under normal denaturing conditions, while Met 80 (6th ligand) is less firmly attached and can be replaced from its position by another histidine, cyanide, imidazole, or azide. The Met 80-heme bond can also be destroyed under very low or very high pH values.

The heme is tightly wrapped in hydrophobic residues. The hydrophobic amino acids such as Phe, Met, Ala, Val, Leu, and Ile arise in groups along the chain. Chains confined by residues 6-20 and 52-74 form right and left channels, respectively. These channels include primarily hydrophobic amino acids. Each channel has at least two aromatic rings aligned almost in parallel. Aromatic groups are markedly essential for the protein functioning as well. Phe 10 and Phe 82, Trp 59, Tyr 48, Tyr 67, and Tyr 74, and are completely invariant among species. The lysine content of the protein is very high (19 residues in the horse cyt c), however, lysines are not evenly spread along the polypeptide chain, but are grouped in clusters. Specifically, each channel described above is surrounded by a cluster of lysine residues, and these sites of the protein are exposed to surface. The lysines are almost invariable among the species. Most of the negatively charged side-chains such as aspartic and glutamic acids are positioned at the top rear of the protein. It is remarkable that the clusters of positively and negatively charged side amino acids are localized exclusively on the surface of the protein. The authors (Fisher et al., 1973) hypothesize that the charged sites on the surface of cyt c provide a platform for the interactions of the protein with other enzymes and small molecules.

Bushnell and coworkers used 1.94 Å resolution X-Ray diffraction technique to refine the structure of horse heart cyt c (Bushnell et al., 1990). The authors also performed a detailed comparative analysis of the structures of horse heart cyt c and the cyt c structures of other eukaryotic organisms. In horse cytochrome *c*, the side-chain of Phe 82 is set close to heme in a parallel position to the heme. This feature of residue 82 was also found in other cyt c proteins, and apparently it does not depend on the oxidation state of cyt c. Five water molecules are located in the fixed sites of cyt c of several organisms. Three of them are positioned on the surface of the protein. One water molecule facilitates the electrostatic interactions between arginine 38 and a heme propionate. The second water molecule is positioned close to the iron atom of the heme and forms hydrogen bonds with Asn 52, Tyr 67 and Thr 78. This water molecule plays a significant role in the enzymatic reactions involving cyt c, as it was shown to shift upon change in the oxidation state of the iron atom in the heme group.

It has been long known that in the folded state of cyt c the fluorescence of an amino-acid residue tryptophan, Trp 59, is quenched because of the Forster energy transfer to the heme group. In a folded state of cyt c, the distance between Trp 59 and the heme moiety averaged over the ensembe is less than 20 Å, while in an unfolded state the same distance was determined to be ~33 Å (Sanchez et al., 2008). Tryptophan fluorescence along with stopped-flow methods, circular dichroism, and NMR c has been intensively utilized in the studies of the folding kinetics of the protein. The folding-kinetics studies showed that the distance between Trp 59 and heme shortens considerably within a millisecond as indicated by fluorescence quenching. Two noticeable kinetic phases with time constants of approximately 20 and 300 ms and a minor phase of 5 s follow the first fast phase. The formation of partially folded intermediates was observed during the early folding steps. The first intermediate forms within the first milliseconds, it has a well-defined helical secondary structure, wavering basis and no steady hydrogen bonds. A second intermediate forms at a time scale of several decades of milliseconds. It has well-defined hydrogen bonds in two helices and the enhanced overall firmness. The aromatic residues in the second intermediate are still loosened and exposed to the external environment rather than contributing to the intermediate's folded structure (Roder 1988; Elöve et al., 1992).

NMR studies of the denatured horse cyt c revealed the presence of a highly protected His 18 amide suggesting that this residue is ligated to the heme in the unfolded state of the protein (Elöve et al., 1994). The folding studies also revealed the existence of kinetic phases, which were ascribed to the dissociation of histidine ligands His 26 and His 33 from the heme in the unfolded state. The ligation of these non-native states to the heme forms partially folded intermediates (Elöve et al., 1994).

Takahashi and coworkers employed resonance Raman scattering to analyze kinetics of cyt c folding (Takahashi et al., 1997). Their experiments shows that the initial folding phase occurs within 0.1 ms. Remarkably, the protein was also found to be entrapped in a misfolded state, in which methionine ligand was substituted with His. This intermediate state slowed down the folding process. In the second phase, one of the histidines, His 26 or His 33, linked to heme was substituted with Met.

Russell and coworkers employed ^{1}H NMR spectroscopy to investigate the ligation of heme in different unfolding intermediates (Russell et al., 2000). Under week denaturing conditions, they detected an unfolding state with Met 80 ligand displaced by a lysine residue. The Gibbs free energy change for the process of the transition between the native cyt c state and lysine-substituted state was determined to be 17 kJ/mol, which is essentially lower than the ΔG for the transition from native to bis-histidine ligation state (46 kJ/mol). Remarkably, the study showed that both intermediate states possess distinct non-native geometry of heme ligation.

In a later study, Maity and coworkers, utilized hydrogen exchange method to characterize the dynamic of cyt c folding to its natural state (Maity et al., 2005). In this approach, hydrogen atoms, protected in the native protein state, exchange with solvent, when the protein transmits through unfolding intermediates. They showed that cyt c consists of five domains. These domains fold by the stepwise association of the building blocks. Alternatively, the unfolding of each of the domain occurs only after the denaturation of the preceding domain. The structure of cyt c with its heme group, axial ligands, Met 80 and His 18, along with His 26, His 33, Trp59, and component foldon units are shown in Figure 1.

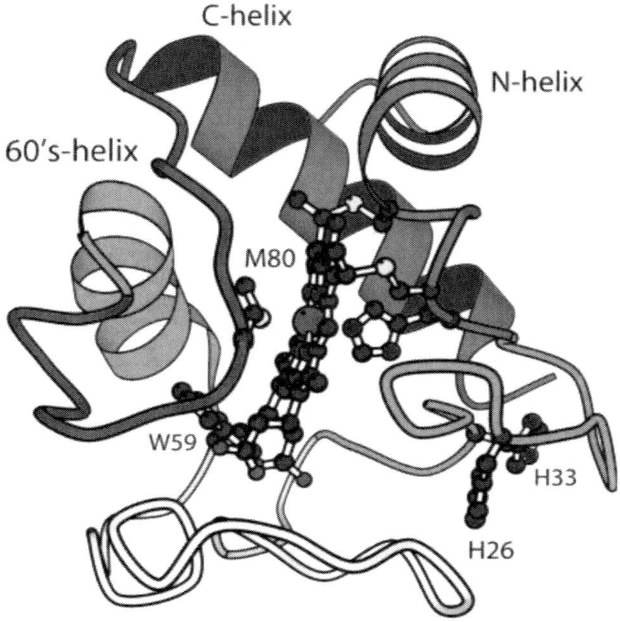

Reprinted with permission from Maity et al., 2005, *P. Natl. Acad. Sci USA 102*, 4741. Copyright (2005) National Academy of Sciences, U.S.A.

Figure 1. The structure of cyt c. The constituting building domains are shown in ribbon diagram. A heme group, a distal axial ligand Met 80, a proximal axial ligand His 18, and three other amino acids Trp 59, His 26, and His 33 are shown in a ball-and-stick model.

2. Interactions of Cyt c with Phospholipids

Cyt c is attached to the IMM and is in close contact with its major constituents phospholipids. Interactions between cyt c and lipids play a crucial role in the modulation of protein's structure, dynamics, and function, specifically processes involving cyt c release from mitochondria and triggering apoptosis.

Pinheiro and coworkers performed a comprehensive study on how lipid vesicles of dioleoyl phosphatidylserine (DOPS, Figure 2A) influence the structure and unfolding kinetic of cyt c (Pinheiro et al., 1997). The stretching of Met 80 to the heme from the crevice of cyt c results in the ligation of the sulfur atom on methionine to the iron atom oh the heme. This axial coordination has a specific absorbance band at 695 nm. The disruption of the Fe-S(Met 80) bond is a signature of the opening of the cyt c crevice. Remarkably, upon cyt c binding to DOPS vesicles the absorption at 695 nm disappeared.

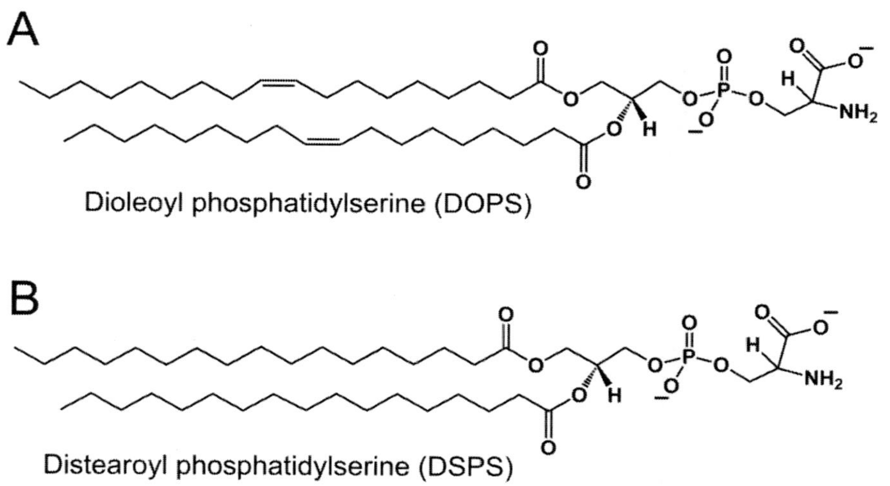

Figure 2. The structures of phosphatidylserine molecules.

The kinetic studies involving the usage of denaturant guanidine hydrochloride (GuHCl) showed that before the structural unfolding can occur the Fe-S(Met 80) bond has to break. The disruption of the 6th axial coordination on the heme thus facilitates the unfolding, and the dissociation of Met 80 from the heme is the rate-limiting step in the whole unfolding mechanism.

The concentration of negatively charged groups on the membrane surface results in the attraction of protons due to the electrostatic forces. Consequently, the local pH significantly decreases. Prats and coworkers performed measurements of the electrostatic surface potential for several phospholipids and showed that the pH of the interfacial water phase in the case of phosphatidylsetine may reach the value of 3.2, which is 3.6 units lower than the pH of the bulk water phase (Prats et al., 1986). The appearance of the low pH media in the vicinity of cyt c heme disrupts the S-Fe bond, which assists the unfolding of the protein. This explains the stimulation and acceleration of cyt c structural unfolding by DOPS vesicles.

The unfolding of cyt c upon binding to DOPS was also supported by the Soret absorbance and tryptophan fluorescence measurements. The 410 nm band indicative of the native folded state of cyt c shifted to 407 nm signifying alterations in the spin of the iron atom caused by the disturbances of the bonding between the heme and its axial ligands, most likely Met 80. The fluorescence intensity of Trp 59 was markedly enhanced upon cyt c binding to DOPS, revealing the increase of the average distance between Trp 59 and heme compared with that in the native protein state.

The circular dichroism (CD) studies performed by Pinheiro and colleagues showed that the compact packing of the core residues in cyt c is disrupted upon the interactions with DOPS vesicles. The CD experiments also revealed that the coupling between π-π^* transitions of the heme and those of the aromatic side-chain amino acids surrounding the heme such as Phe, Tyr, and Trp is destroyed in the presence of DOPS, indicating the disturbance of the tertiary structure of the protein.

Rytomaa and Kinnunen employed fluorescence spectroscopy to study the mechanism of cyt c binding to liposomes containing PG and CL (Rytomaa and Kinnunen, 1994; Rytomaa and Kinnunen, 1995). The authors found that the binding of the protein to these phospholipids is irreversible. A mechanistic model was proposed for the protein-phospholipid interactions at the membrane surface. According to this model, two acyl chains of a phospholipid stretch out in opposite directions from the phosphate group forming an extended lipid conformation. This is supported by NMR and X-crystallography data, which show that the torsion angle between atoms C-2 and C-3 of the main glycerol group is in the antiperiplanar range (Hauser et al., 1988). In this configuration, one of the acyl chains penetrates the hydrophobic channel in the protein, while the other chain rests in the membrane. Rytomaa and Kinnunen suggested there are two well-defined acidic lipid binding sites on the surface of cyt c: the A-site and the C-site. The A-site binds phospholipid molecules with deprotonated phosphate. This binding occurs via the electrostatic attraction of the negatively charged phosphate and the positively charged conserved Lys 72 and Lys 73. The protein-lipid interaction at the C-site occurs via hydrogen-bonding between the protonated phosphate of the lipid and the conserved Asn 52 residue. It is noteworthy that since the structure of CL involves two phosphate groups and consequently four acyl chains, CL may bind cyt c via two mechanisms at the same time.

In a later study, Tuominen and coworkers provided an additional proof for the extended conformation of phospholipid bound to cyt c (Tuominen et al., 2002). The authors employed a modified cyt c in which iron was substituted with Zn^{2+}. A heme group with coordinated Zn^{2+} is inherently fluorescent, excitation and emission wavelengths being 412 nm and 588 nm, respectively. Variations in the fluorescence intensity may provide information about the processes occurring in the vicinity of the heme. Brominated lipid derivatives

(9,10)-dibromostearate and 1-palmitoyl-2-(9,10)-dibromo-*sn*-glycero-3-phospho-*rac*-glycerol ($P(Br_2)SPG$) were used as a model to investigate mechanisms of the interactions between cyt c and phospholipids. When in contact with a fluorescent molecule, bromine atoms quench fluorescence. Monitoring fluorescence of heme-Zn group can thus provide information about the direct contacts between the heme and the phospholipid molecule. The experiments showed that ($P(Br_2)SPG$) quenches fluorescence of the Zn-substituted heme. The quenching saturates at cyt c:PG molar ratio of approximately 1:40 (Tuominen et al., 2002). This result provides direct evidence of the interactions between PG and heme, which is located inside the protein.

3. Peroxidase Activity of Cyt c and its Complex with CL

The conformational changes in cyt c caused by the interactions with phospholipids influence its function in mitochondria. The major role of cyt c is the shuttle of electrons between complexes III and IV of the respiratory chain. This shuttle occurs through the changes of the oxidation state of the iron atom of the protein's heme group. When a phospholipid binds cyt c via electrostatic interaction and through the insertion of the acyl chain into the protein's hydrophobic channel, the structure of the protein becomes distorted. This distortion lowers the energy barrier for H_2O_2 molecule or another organic peroxide molecule to reach the heme and substitute the Met 80 ligand. The reduction of the energy barrier for protein-peroxide interactions ultimately leads to the activation of protein's peroxidase function. The peroxidase activity of cyt c allows the protein to perform other essential functions in mitochondria in addition to electron transfer in the oxidative phosphorylation process.

Kagan and coworkers employed chemiluminescence, fluorescence, and electron paramagnetic resonance (EPR) techniques to demonstrate that cyt c-CL complex possesses a strong peroxidase activity (Kagan et al., 2005). The authors observed an EPR signal in a sample of cyt c bound to tetraoleoyl CL (TOCL) and tetralinoleoyl CL (TLCL) upon the addition of H_2O_2 (Figure 3A and 3B). The spectroscopic features of the signal revealed that the presence of a radical associated with the protein, which typically forms under the reactions of hemoproteins with peroxides. The reaction between the native ferric heme state (Fe^{3+}) and hydroperoxides results in two-electron oxidation of the heme and the formation of Compound I, or a ferryloxoprotoporyphrin cation radical. This intermediate state has an iron atom in its oxoferryl state (Fe^{4+}=O) and a radical cation located on the π-system of the heme. Compound I is reduced to Compound II via one-electron reduction process. In Compound II, the iron atom is in its ferric state and a Tyr residue is oxidized to a tyrosyl radical. It is believed that Compound II possesses activated oxygenase functionality. Tyrosyl radical of Compound II abstracts a hydrogen atom from a lipid molecule forming a carbon-centered radical on the lipid and thus facilitating the formation of lipid peroxide through the addition of molecular oxygen and finally the formation of lipid hydroperoxide. Compound II is reduced to the native state through one-electron reduction. The catalytic reaction scheme of cyt c active center is shown in Figure 4. The tyrosyl radicals have been detected in many hydroperoxides and their EPR spectra have been well characterized. These enzymes include soybean leghemoglobin, human hemoglobin, human myoglobin, sperm whale myoglobin, and horse myoglobin (Svistunenko, 2005). Kagan and coworkers also confirmed the formation of tyrosyl radicals in cyt c-TOCL and cyt c-TLCL systems by spin trap experiments using 2-methyl-2-nitrosopropane (MNP). The authors detected MNP-tyrosyl radical adducts and dityrosine cross-linked products. It is noteworthy that only the complexation of cyt c with unsaturated CL such as TOCL and TLCL resulted in the detectable peroxidase activity of the protein. When cyt c was in a free state, bound to the saturated tetramyristoyl CL (TMCL, Figure 3C), or dioleoyl phosphocholine (DOPC) no EPR signal was detected (Kagan et al., 2005). This observation emphasizes an exclusive role of phospholipid CL, specifically CL having unsaturated acyl chains, in the peroxidase activity of cyt c.

In a later study, Kapralov and coworkers investigated the involvement of four different Tyr residues of cyt c in the peroxidadtion of unsaturated CL (Kapralov et al., 2011). The authors studies how the peroxidase activity of cyt c-CL is affected by the following mutations: Tyr48Phe, Tyr97Phe, Tyr67Phe, and Tyr74Phe. The data showed that the mutation at the 67th position resulted in the greatest inhibition of the peroxidase catalysis. It was also shown that Tyr 67 is located in close proximity to the heme of cyt c and to the distal axial ligand Met 80. The authors concluded that Tyr 67 is a likely participant in the electron transfer reaction occurring on the active center during the peroxidase catalysis (Figure 4).

Belikova and coworkers investigated the relationship between the peroxidase activity of cyt c-CL complex and the conformational transitions in the protein (Belikova et al., 2006). The authors correlated the measurements of absorbance at 695 nm, characteristic of the Fe-S(Met 80) bond, with the EPR studies. After the incubation of cyt c with CL/PC liposomes at a low ionic strength the absorbance intensity at 695 nm decreased significantly, indicating the disruption of Fe-S(Met 80) bond. The addition of KCN to the solution of cyt c already bound to CL resulted in the additional decrease of the absorbance at 695 nm by 70%, revealing a low potential barrier in the reaction of Met 80 substitution by the cyanide.

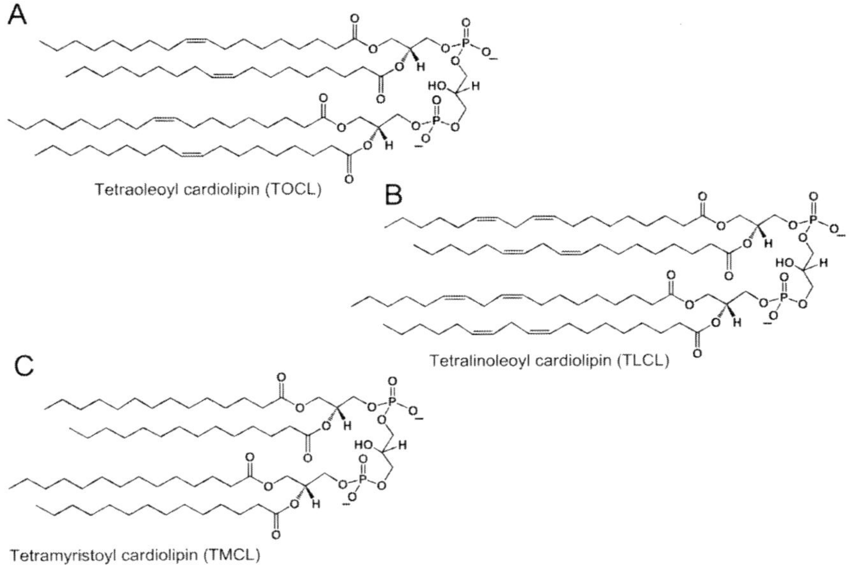

Figure 3. The structures of cardiolipin molecules.

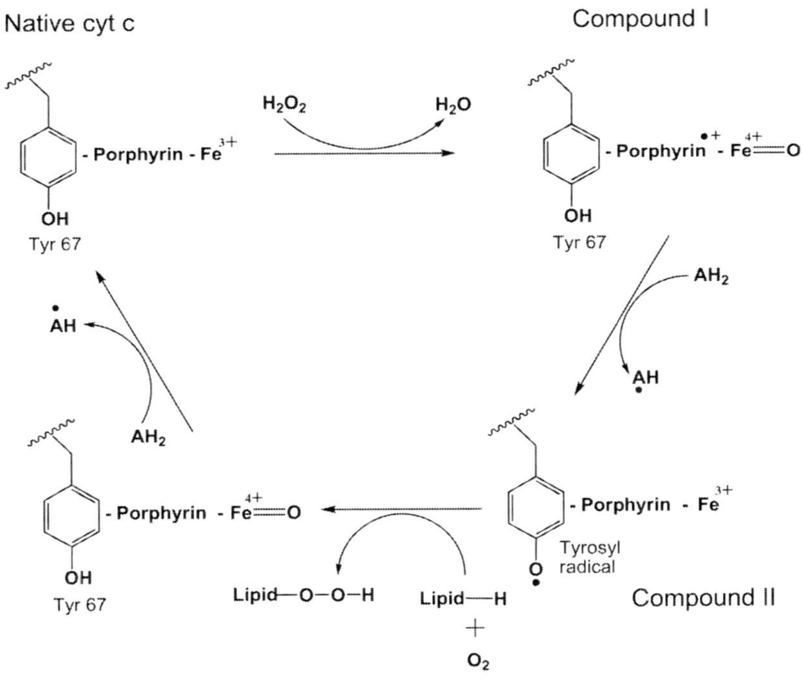

Figure 4. The catalytic scheme of cyt c involving peroxidase steps and an oxygenase step. Cyt c in its native ferric state reacts with hydrogen peroxide and forms ferryloxoprotoporyphrin cation radical (Compound I), with is a product of two-electron oxidation of the native enzyme. A reaction of one-electron reduction converts Compound I into Compound II, in which Tyr residue is oxidized to tyrosyl radical and iron is in its ferric state. This intermediate state with a radical located on Tyr residue plays a role of oxygenase and catalyzes the addition of oxygen molecule to a lipid molecule with the ultimate formation of lipid hydroperoxide. A second reaction of one-electron reduction converts Compound II into its native state.

The addition of the same amount of KCN to the free cyt c only led to 10% decrease of the absorbance characteristic of Met 80 ligation, indicating the strong binding of the distal axial ligand in the intact protein compared with that in CL-bound protein. Remarkably, no decrease of absorbance at 695 nm was detected upon the addition of CL to cyt c in a high ionic strength environment.

To quantify the peroxidase activity of the cyt c-CL complex, the authors utilized the reaction of the catalytic oxidation of etoposide to etoposide phenoxyl radical by H_2O_2. The analysis of the radicals' EPR spectra showed that the production of etoposide radicals increased significantly upon the addition of CL to cyt c, revealing the enhancement of the peroxidase activity of CL-bound cyt c compared to that of the free protein. However, while the absorbance at 695 nm did not change within 30 min after the addition of a certain amount of CL to cyt c, the peroxidase activity of the protein increased drastically at the same conditions. This observation revealed the fact that the catalytic activity of the CL-bound protein is not directly proportional to the weakness of the Fe-S(Met 80) bond. The protein gains a significant peroxidase activity without the disruption of the bond between the heme and the 6th ligand (Belikova et al., 2006). The authors also correlated the results of the EPR experiments with the studies of tryptophan fluorescence as a measure of cyt c unfolding degree. They found that the enhancement of the peroxidase activity of cyt c occurs much faster than the unfolding of the protein.

The peroxidase activity of cyt c increases drastically upon binding to CL; this indicates the importance of the phospholipid binding step in triggering the catalytic function of the protein. However, it was shown that the peroxidase activation already occurs when cyt c is bound to a phospholipid weakly, through electrostatic interactions, without any significant changes in the tertiary structure of the protein and without the disruption of the Fe-S(Met 80) bond (Belikova et al., 2006). The authors justify their observations by referring to the structural analysis of cyt c, which demonstrates the presence of small opening in the protein formed by Lys 79, Met 80, Ile 81, and the heme. The interactions of the protein with anionic phospholipids may result in the loosening of the structure of this opening and simplifying the access of H_2O_2 to the heme, where it may replace Met 80 and react with the iron of the heme thus initiating the catalytic process.

The authors also assessed the cyt c-phospholipid association constants using acridine 10-nonyl bromide (NAO) in a competitive binding assay (Belikova et al., 2006). NAO is a fluorescent dye; it is used as a highly specific probe for CL as it binds specifically to this phospholipid through both hydrophobic and electrostatic interactions. It was found that the binding strength of cyt c to phospholipids increases with the number of double bonds in acyl chains, the lowest affinity constant being with TMCL, a phospholipid with saturate acyl chains: TMCL < TOCL = TLCL < bovine heart CL. Tryptophan fluorescence measurements showed that the ability of TOCL to unfold cyt c is much greater than that of TMCL. Remarkably, the peroxidase activity of cyt c-CL complex evaluated in the same work had similar trend within different CLs. Thus, the activity of cyt c-TOCL complex (mono-unsaturated phospholipid) was five times greater than that of cyt c-TMCL complex (saturated phospholipid). The experiments revealed that the protein-phospholipid binding strength is correlated with the peroxidase activity, indicating that the mechanism of protein-lipid binding and the type of biochemical interactions formed in the complex are crucial in initiating the catalytic activity of the hemoprotein.

One of the most important findings of Belikova and colleagues was that in the high ionic strength buffer containing 1 M of KCl the spectrum associated with the Fe-S(Met 80) bond remained unchanged upon the addition of the phospholipids to cyt c (Belikova et al., 2006). The peroxidase activity of phospholipid-bound cyt c remained at the same level as that of the free protein in the media containing 1 M KCl. This observation underscores the importance of the electrostatic binding in cyt c-phospholipid interactions. Stepanov and colleagues investigated the mechanism of the anionic interactions between cyt c and phospholipids by combining surface plasmon resonance (SPR), fluorescent, and chemiluminescent techniques (Stepanov et al., 2009). The following phospholipids were investigated: DOPS (Figure 2A), distearoyl phosphatidylserine (DSPS, Figure 2B), TOCL (Figure 3A), TMCL (Figure 3C), DOPA (Figure 5A), and DSPA (Figure 5B). The authors measured the protein-lipid association constants using the SPR technique. The values of the constants ranged between 2×10^3 M^{-1} and 2×10^4 M^{-1}, the lowest constants being that of PA and the highest of CL. This observation could be explained by the presence of four acyl chains that insert into cyt c cavity (Rytomaa and Kinnunen, 1995). The authors did not observe the difference in the cyt c association strength for phospholipids having different saturation degree of their acyl chains. The binding constants varied depending on the characteristic group of phospholipids. Notably, the values of the cyt c-lipids association constants decreased by two orders of magnitude at high salt concentrations compared to those measured at low ionic strength, as determined by SPR experiments. This again suggests the crucial role of electrostatic forces in the interactions between the protein and phospholipids (Stepanov et al., 2009).

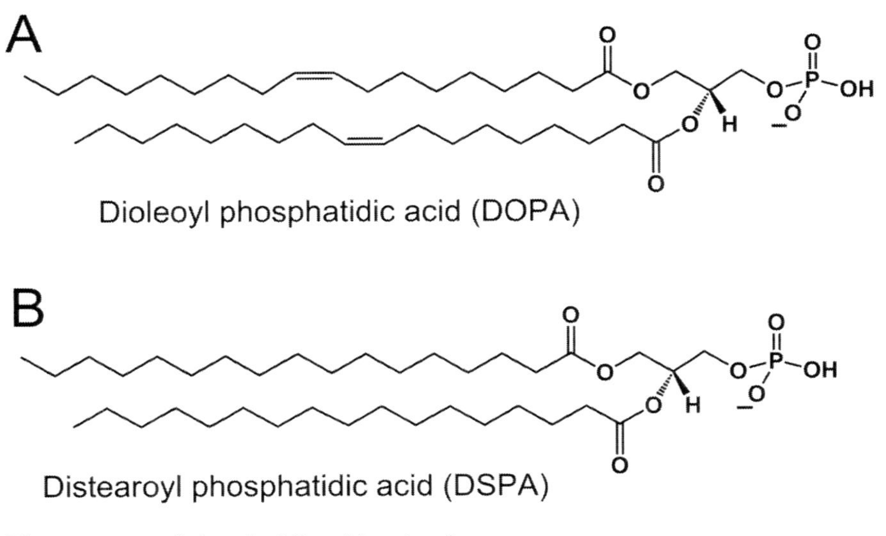

Figure 5. The structures of phosphatidic acid molecules.

Using Trp 59 fluorescence, the authors studies how the phospholipids having different specific groups and different saturation degree of acyl chains affect the unfolding of cyt c. The results showed that the effect of phospholipids with saturated fatty acids on cyt c unfolding state was almost negligible. On the other hand all types of phospholipids with unsaturated fatty acids caused dramatic increase of Trp 59 fluorescence, indicating significant unfolding of cyt c. The effect on the protein unfolding increased in the following order: DOPS < DOPA < TOCL. Remarkably, the measurements of cyt c peroxidase activity in a phospholipid-bound state revealed trends similar to those observed in protein unfolding studies. Namely, no increase in cyt c peroxidase activity was detected upon its binding to saturated phospholipids. However, considerable increase in the catalytic activity was observed for complexes of cyt c with unsaturated phospholipids, the activity trend being the following: DOPS < TOCL < DOPA (Stepanov et al., 2009).

Recently Hanske and coworkers employed time-resolved FRET to study the properties of cyt c in its CL bound state (Hanske et al., 2012). They revealed that the electrostatic interactions between CL and cyt c stimulate drastic unfolding of the protein. However, the degree of protein unfolding varies between polypeptides. The authors stressed the importance of this conformational diversity for the peroxidase function of the protein and hypothesize that the peroxidase activity is dominated by largely open structures due to the highest heme accessibility (Hanske et al., 2012).

Phospholipid Hydroperoxide Glutathione Peroxidase (Gpx4) as an Inhibitor of Apoptosis

1. The Catalytic Cycle and Structure of GPx4 and Other GPx-Isoforms

The major role of glutathione peroxidases (GPx) is the protection of cells from oxidative peroxides generated internally as a result of metabolic oxidation reactions and other catalytic processes associated with normal cellular life cycle (Arai et al., 1999; Muller et al., 2007). GPx is a selenoenzyme that catalyzes the reduction of hydrogen peroxides and more complicated organic peroxides by glutathione (GSH). The key amino acid in the catalytic cycle of GPx is selenocysteine (Sec), an analog of cysteine having sulfur atom substituted by selenium atom. The selenol (E-SeH) is oxidized by peroxide to selenenic acid (E-SeOH). E-SeOH then reacts with the GSH to form selenenyl sulfide product (E-SeGH). E-SeGH reacts with the second GSH molecule producing oxidized glutathione (GSSH) and the enzyme in its native catalytic form, E-SeH (Flohe et al., 1972).

The catalytic scheme of GPx is shown in Figure 6. Under high concentration of peroxide, the enzyme may be oxidized to seleninic acid (E-SeO$_2$H) as a secondary product, but it is believed that this pathway does not contribute to the major catalytic process and is off the catalytic cycle of GPx. The presence of selenolate anion (E-S$^-$), the oxidized enzyme seleninic acid, and the selenosulfide bond in E-SeGH were confirmed by crystallographic studies (Epp et al., 1983).

In mammalians, eight different GPx isoforms encoded by different genes have been identified, GPx1-8. Among these isoforms GPx4 is exceptional due to its ability to reduce complex lipid hydroperoxides. Its major function, exclusive among antioxidant enzymes, is the protection of membranes from oxidation. Recently, a neuroprotective role of GPx4 was also demonstrated (Bellinger et al., 2011). Another peculiar property of GPx4 is its tendency for polymerization, which is not typical for other GPx isoforms.

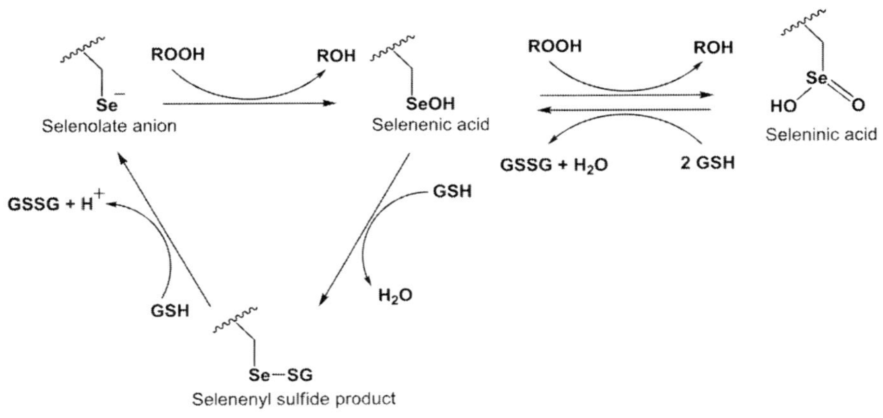

Figure 6. The general catalytic scheme of glutathione peroxidase. A selenolate anion, a deprotonated product of Sec 46, reacts with hydroperoxide forming selenic acid. Selenic acid reacts with GSH producing selenyl sulfide product, which in turn reacts with another GSH molecule producing oxidized GSH, GSSH, and a Sec residue in its catalytically active form, selenolate anion. Selenic acid can also be oxidized to seleninic acid via the oxidation by hydroperoxide, the reaction representing a minor secondary product of the major catalytic cycle.

There are three different variants of GPx4: cytosolic isoform c-GPx4, nuclear isoform n-GPx4, and mitochondrial isoform m-GPx4. There are no direct structural data on m-GPx4. Here we will focus on the description of the crystallographic data on c-GPx4. c-GPx4 is a small protein with a molar weight of 19.5 kDa. The analysis of GPx sequences revealed the conservation of three amino acids: Sec 46, Gln 81, and Trp 136. These residues are catalytically active as indicated by mutagenesis analysis. Interestingly, c-GPx4 can also utilize protein thiols as electron donors in a catalytic cycle in addition to GSH. Most of the structural information on GPx4 is based on the crystallographic data of other GPx-isoforms. For a long time, the direct structural data on GPx4 were not available, the principle reason being the deficiency of an effectual biological source of this peroxidase. Another reason for the lack of the structural information on GPx4 is the difficulty associated with the expression of proteins containing selenocysteine in recombinant systems.

Scheerer and coworkers overexpressed the Sec46Cys mutant of human c-GPx4 in *E. coli* and performed crystallographic studies of the protein at a resolution of 1.55 Å in order to elucidate the basis of its peculiar catalytic activity and to explain the mechanism of its polymerization (Scheerer et al., 2007). The data indicate that GPx4 is a monomeric protein consisting of 170 amino acids. A structural model of the protein represents a typical thioredoxin pattern composed of four α-helices and seven β-strands. α-helices are positioned at the protein surface. Five of the seven β-strands are grouped in a central β-sheet (Figure 7).

The data collected by Scheerer and colleagues support the assumption that the major catalytic amino acid, Sec 46, is in a deprotonated form (Figure 6), while the other two amino acids of the catalytic triad Gln 81 and Trp 136 located within the hydrogen-bond distance from Sec 46 form hydrogen bonds with the selenolate anion. The structural element containing the catalytic triad is lined with hydrophobic residues which may facilitate the access of lipid hydroperoxides to the catalytic center during the reaction of the peroxide reduction (Scheerer et al., 2007).

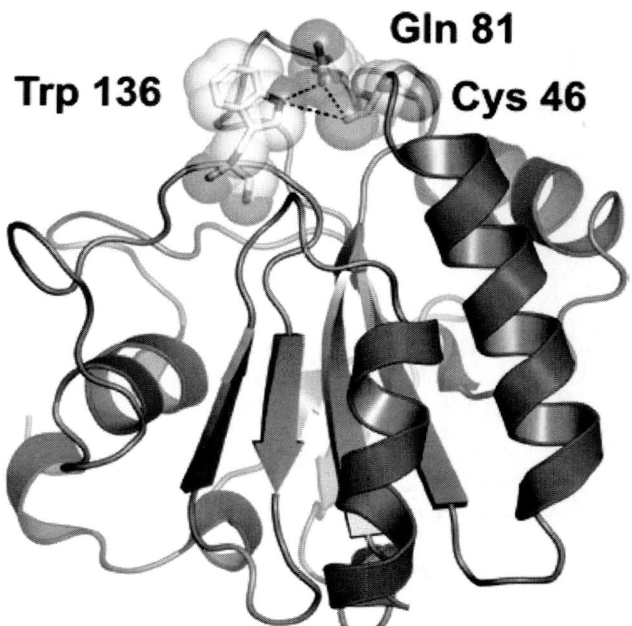

Reprinted with permission from Scheerer et al., 2007, *Biochemistry 46*, 9041. Copyright (2007) American Chemical Society.

Figure 7. The structure of the cytosolic GPx4 enzyme. The ribbon diagram of the backbone conformation shows seven β-strands and four α-helices. The catalytic triad consisting of Cys 46, Trp 136, and Gln 81 are shown in a ball-and-stick model.

The authors compared the protein sequences for various GPx-isoforms and found that GPx4 unlike other isoforms lacks 20 internal amino acids (Scheerer et al., 2007). The crystallography data of the GPx enzymes show that these 20 residues form a surface-exposed loop 1, which covers Trp 136, a constituent of the catalytic triad. This finding strongly suggests that the absence of this loop in GPx4 is designed by nature to facilitate the access of the oxidized lipids to the active site of the enzyme, which is not so easily accessible in other GPx-isoforms. Another uniqueness of GPx4-isoform is the lack of six amino acids forming loop 2, with is present in the other isoforms. The deficiency of this loop in GPx4 leads to the displacement of helix 2. The two factors combined together, namely the appropriately positioned helix 2 and loop 1 mentioned above are parts of the monomer interface in oligomeric GPx-isoforms (Maiorino et al., 2007).

The modification of these structural features in GPx4 results in the monomeric character of its structure, which facilitates the accessibility of the active site in catalytic reactions.

The authors (Scheerer et al., 2007) employed mutagenesis approach, in which various Cys residues were mutated to Ala, to investigate the structural basis for the polymerization tendency specific to GPx4. The following mutants demonstrated a reduced polymerization activity: Sec46Ala, Cys10Ala+Sec46Cys, and Sec46Cys+Cys66Ala. Remarkably, polymerization was not inhibited when surface exposed cysteines, namely Cys 107 and Cys 148, were mutated to Ala. The crystallographic data also indicate that Sec 46 of one GPx4 monomer approaches Cys 10 and Cys 66 of other monomers suggesting the formation of mixed selenocysteine-cysteine bonds and 3D polymer networks of GPx4.

2. The Functions of m-GPx4 and its Anti-Apoptotic Dynamics

Various studies revealed that m-GPx4 is a multifunctional protein. Imai and coworkers measured the expression level of m-GPx4 in human testis and in spermatozoa from fertile and infertile men (Imai et al., 2001). They found that the expression of GPx4 in men diagnosed with oligoasthenozoospermia was significantly suppressed compared with that in healthy men. The authors also employed electron microscopy to perform ultrastructural analysis of mitochondria and found that the morphology of mitochondria in GPx4-deficient cells was anomalous. The results strongly suggest the essential role of m-GPx4 in spermatogenesis. In a later study, another group of scientists showed that the deletion of m-GPx4 causes male infertility as revealed by the reduced sperm quality and structural abnormalities of spermatozoa (Schneider et al., 2009). This result also suggests the vital role of selenium in mammalian male fertility.

Dabkowski and coworkers proposed the usage of m-GPx4 as a therapeutic protectant that could suppress cardiac dysfunction caused by ischemia-reperfusion injury (Dabkowski et al., 2008). They demonstrated that (1) GPx4 overexpressed in mice heart mitochondria diminished the level of lipid peroxidation, and (2) m-GPx4 enhanced the activity of electron transport chain complexes in subsarcolemmal and interfibrillar mitochondria.

Nomura and coworkers investigated the role of GPx4 in mitochondrial apoptosis, specifically they studies how the expression of m-GPx4 affects cyt c-CL binding, CL peroxidation, and cyt c release from mitochondria (Nomura et al., 2000). First, the authors performed *in vitro* studies, in which they showed that cyt c associated with liposomes containing CL, but did not bind to liposomes containing peroxidized CL (CL-OOH), even when excess of these liposomes was added. However, after the liposomes had been directly reduced with GPx4, cyt c did attach to the liposomes. Interestingly, cyt c was able to bind the reduced form of CL-OOH, CL-OH. This result confirmed the significance of the role GPx4 plays in the dynamics of cyt c-CL interactions.

In a second set of experiments, the authors used RBL2H3 cell lines that stably overexpressed m-GPx4 (M15 cells) and control S1 cells in apoptosis assays. The cells were exposed to 2-deoxyglucose (2DG), the inducer of hypoglycaemia and consequently apoptosis. The exposure of S1 cells to 2DG resulted in the (1) substantial elevation of the level of hydroperoxides in 2 hours after the treatment, (2) the release of cyt c from mitochondria following 4 hours after the exposure, and (3) the activation of caspase 3 within 6 hours after the treatment with 2DG. Remarkably, none of these effects was detected in M15 cells following the exposure to 2DG, suggesting the protective effect of m-GPx4 against apoptosis. Interestingly, when M15 cells were exposed to diethyl malate (DEM) that suppresses the intracellular concentration of GSH and thus reduces the catalytic activity of m-GPx4, the cells became sensitive to apoptosis following the 2DG treatment. This result underlines the importance of m-GPx4 in the regulation of the mitochondrial apoptosis (Nomura et al., 2000).

In an alternative *in vivo* experiment, Nomura and coworkers studies the fluorescence of NAO attached to CL in M15 and S1 cells. As a result of NAO-CL binding mitochondria appeared as yellow particles under the fluorescence microscope. Since the affinity of NAO to CL-OOH is fairly weak, the disappearance of the cell staining would indicate the peroxidation of CL. When S1 cells were exposed to 2DG, the yellow color faded within the first two hours. However, when M15 cells were treated with 2DG, almost no staining change was observed. Thus, the overexpression of m-GPx4 prevented the formation of CL-OOH in mitochondria.

The experimental data combined together suggest that the formation of CL-OOH in mitochondria is the key event in the initiation of apoptosis and that m-GPx4 functions as an apoptotic factor by suppressing the production of CL-OOH (Nomura et al., 2000).

In a different study, Nomura and colleagues (Nomura et al., 1999) studied L9 cells overexpressing non-mitochondrial GPx4 in addition to M15 cells overexpressing mitochondrial GPx4 and the control cells S1. The researchers observed cyt c release, caspase 3 activation, and apoptosis in response to 2DG treatment in S1 and L9 cells, but not M15 cells. The level of lipid peroxides was drastically increased in mitochondria of S1 and L9 cells after 2DG exposure, however, the level of peroxides in mitochondria of M15 cells was essentially lower. This result indicates that only mitochondrial GPx4 inhibits apoptosis conveyed by mitochondrial death pathways, while non-mitochondrial GPx4 remains inactive. Interestingly, m-GPx4 also suppressed apoptosis stimulated by UV radiation, cycloheximide, actinimycin D, etoposide, and stautosporine. The overexpression of m-GPx4 did not change the expression level of the proteins from Bcl-2 family. m-GPx4 did not protect cells from apoptosis induced by Fas-specific antibodies (Nomura et al., 1999). These observations altogether stress (1) the importance of lipid peroxide in triggering apoptosis, (2) the crucial role of m-GPx4 in suppressing apoptosis specifically by interfering with the processes of cyt c-CL interactions.

The Role of Peroxidized Lipids in the Permeabilization of the Mitochondrial Membranes

Kagan and coworkers showed that the cyt c-CL complex selectively oxidizes CL in mitochondrial membrane during apoptosis (Kagan et al., 2005). Remarkably, even though the acyl chains of other major phospholipids PE and PC have a greater propensity to peroxidation than the acyl chains of CL, it is only CL that is oxidized to CL-OOH. The authors also demonstrated that the formation of CL-OOH occurs at the early stages of apoptosis and that CL-OOH specifically stimulates the release of proapoptotic factors from the mitochondrial intermembrane space. This finding along with the knowledge that m-GPx4 inhibits apoptosis by suppressing the peroxidation level of CL suggest that of CL-OOH plays a vital role in triggering mitochondrial apoptosis. CL binds to many different proteins that coordinate mitochondrial apoptosis including cyt c, Bcl-2 proteins, and adenine nucleotide translocase (ANT). The appearance of hydroperoxide groups on CL alters the strength and the nature of these interactions thus altering the overall mitochondrial state and prompting PTP opening followed by the release of the proapoptotic factors into the cytoplasm where they activate caspases eventually causing cellular death.

Biochemical events following the formation of CL-OOH and triggering PTP opening and apoptosis are not comprehensively understood, however, they have been intensively studied, and a number of models explaining the observed correlation between the appearance of CL-OOH and the stimulation of apoptosis have been proposed. Petrosillo and coworkers studied the interaction of CL-OOH with mitochondrial membranes of rat heart (Petrosillo et al., 2006). The authors treated mitochondria with micromolar concentrations of CL-OOH and observed mitochondrial matrix swelling, the collapse of intermembrane potential, and the release of pre-accumulated Ca^{2+} and cyt c. When mitochondria were treated with CSA or bongkrekic acid (BA) along with CL-OOH these events were significantly inhibited. BA is a well-known inhibitor of ANT. The supression of cyt c release observed in the presence of this ligand suggests the involvement of ANT in PTP opening. The data also indicated that CL-OOH lowered the maximum level of Ca^{2+} required for triggering PTP opening. Overall the results of this study are consistent with a two-step model. According to this model, CL-OOH is required for (1) cyt c dissociation from the outer surface of the IMM and (2) PTP opening. These two steps are coordinated and together lead to cyt c release from mitochondria (Petrosillo et al., 2006).

ANT is a homodimeric protein located in the IMM. Its major function is the exchange of mitochondrial ATP for cytosolic ADP. Numerous studies have also indicated the involvement of this protein in the opening of a large nonspecific pore, which is also called PTP. The crystallography data show the presence of three CL molecules per ANT monomer. These molecules are tightly attached to the protein, two molecules being sandwiched between the monomers on the matrix side (Nury et al., 2005). Nury and colleagues hypothesized that CL molecules located at the ANT interface control the interactions between the monomers and also regulate the overall conformation of the protein. The peroxidation of CL results in a conformational change that facilitates PTP opening.

Brustovetsky and Klingenberg proposed a model which explains how Ca^{2+} cations together with CL molecules modulate conformations of ANT. According to this model, Ca^{2+} cations bind to the negatively charged headgroups of CL thus competing with the lysine residues of the protein. The positive charges accumulate at the ANT interface causing intramolecular repulsion and gate opening (Brustovetsky and Klingenberg, 1996). Crystallography data indicate that aromatic residues of ANT such as Trp, Tyr, and Phe stack to acyl chains of CL (Nury et al., 2005). It may be suggested that the peroxidation of CL could result in the weakening of the CL-ANT binding, which in turn would lead to partial release of lysine residues from CL headgroups and pre-accumulation of these positively charged residues at the protein interface. In this situation, the concentration of Ca^{2+} cations required to displace lysine residues from CL phosphate groups would be lower, which would explain why the appearance of CL-OOH lowers the threshold of Ca^{2+} necessary for PTP opening.

Imai and coworkers demonstrated that the activity of ANT is inhibited by CLOOH in liposomes (Imai et al., 2003). Remarkably, this activity was recovered under the incubation of liposomes with m-GPx4, which reduced CL-OOH to CL-OH. The authors also showed that the overexpression of m-GPx4 in cells subject to hypoglycaemia precluded the inactivation of ANT and PTP opening. The enzyme also prevented the release of cyt c and AIF from mitochondria.

This result suggests an important role of m-GPx4 in the modulation of the opening of the non-specific mitochondrial pore and apoptosis. The experimental results and the proposed models overall suggest the importance of the catalytic activities of both m-GPx4 and cyt c-CL complex in coordinating the initial phases of apoptosis. The relationships between the functions and activities of these peroxidases need to be further explored.

The peroxidation of CL also influences the permeabilization of the OMM. The oligomerization of Bax and Bak in the OMM leads to the formation of channels. These channels allow the transportation of cyt c through the OMM into the cytoplasm (Dejean et al., 2006). Jiang and coworkers showed that there is correlation between the processes of CL peroxidation, Bax/Bak control of ROS generation, and the transformation of cyt c into CL-specific peroxidase (Jiang et al., 2008). Mari and coworkers employed liposomes to mimic the composition of OMM and demonstrated that the insertion of CL-OOH into the bilayer increases the membrane-permeabilizing potential of Bax. This effect is due to the remodeling of the lipid-bilayer structure via the lamellar-to-inverted hexagonal lipid-phase transition (Mari et al., 2008). In this view, CL-OOH plays a dual role in triggering mitochondrial apoptosis. First, it facilitates the detachment of cyt c from the IMM, and second, it facilitates the formation and expansion of the channels in the OMM allowing the passage of cyt c through the membrane.

Conclusion

Mitochondrial peroxidases regulate the peroxidation level of phospholipids, thus controlling the biophysical properties of membranes and coordinating apoptosis. The major phospholipid selectively peroxidized in the process triggering mitochondrial apoptosis is CL. While m-GPx4 suppresses the peroxidation level of CL, cyt c activates into peroxidase via binding to CL and catalyzes the formation of hydroperoxides on CL. CL-OOH then facilitates the dissociation of cyt c from the IMM and stimulates the formation of pores in the membranes allowing the escape of cyt c from mitochondria to the cytoplasm. A second-order rate constant of GPx4 enzyme with a phospholipid hydroperoxide used as a substrate was determined to be of the order of 10^6 M^{-1} s^{-1} (Sunde and Hadley, 2010).

The value of the second-order rate constant of free cyt c is only 1 M^{-1} s^{-1}. However, at a high lipid-protein ratio and a low ionic strength, the activity of cyt c-CL complex can be as high as 50 M^{-1} s^{-1} (Belikova et al., 2006). In an alternative study, the catalytic activity of cyt c turned into a peroxidase was estimated to be 200 M^{-1} s^{-1} (Kagan et al., 2005). Thus cyt c bound to CL may act as a competitive peroxidase in mitochondria. Huang and coworkers demonstrated that the depletion of CL leads to the enhancement of cellular resistivity to apoptosis. It was shown that the suppression of apoptosis in this case was caused by the inability of cyt c to efficiently associate with CL and to turn into a peroxidase with activity sufficient for the production of CL-OOH (Huang et al., 2008). The authors suggested that by regulating the activity of CL synthase with inhibiting agents the level of CL could be controlled, and this approach could be utilized as an antiapoptotic therapeutic strategy. On the other hand, it was proposed that m-GPx4 could be used as a therapeutic agent that would protect cardiac contractile and mitochondrial functions following ischemia/reperfusion (Dabkowski et al., 2008).

The regulation of the level of m-GPx4 and the control of the peroxidase activity of cyt c-CL complex via CL biosynthesis could be investigated as a complex strategy in treating cancerous and neurodegenerative diseases. Novel strategies have to be developed that would involve the simultaneous determination of the activities of cyt c-CL and GPx4 in order to (1) comprehensively understand the interplay between the roles of these peroxidases (2) to elaborate effective approaches for the regulation of the mitochondrial death pathways.

References

Arai, M., Imai, H., Koumura, T., Yoshida, M., Emoto, K., Umeda, M., Chiba, N., Nakagawa, Y. (1999) Mitochondrial phospholipid hydroperoxide glutathione peroxidase plays a major role in preventing oxidative injury to cells. *J. Biol. Chem. 274*, 4924-4933.

Baines, C. P., Kaiser, R. A., Purcell, N. H., Blair, N. S., Osinska, H., Hambleton, M. A., Brunskill, E. W., Sayen, M. R., Gottlieb, R. A., Dorn, G. W., Robbins, J., Molkentin, J. D. (2005) Loss of cyclophilin D reveals a critical role for mitochondrial permeability transition in cell death. *Nature 434*, 658-662.

Belikova, N. A., Vladimirov, Y. A., Osipov, A. N., Kapralov, A. A., Tyurin, V. A., Potapovich, M. V., Basova, L. V., Peterson, J., Kurnikov, I. V., Kagan, V. E. (2006) Peroxidase activity and structural transitions of cytochrome c bound to cardiolipin-containing membranes. *Biochemistry 45*, 4998-5009.

Bellinger, F. P., Bellinger, M. T., Seale, L. A., Takemoto, A. S., Raman, A. V., Miki, T., Manning-Boğ, A. B., Berry, M. J., White, L. R., Ross, G. W. (2011) Glutathione Peroxidase 4 is associated with Neuromelanin in Substantia Nigra and Dystrophic Axons in Putamen of Parkinson's brain. *Mol. Neurodegener. 6*, 8-16.

Brustovetsky, N., Klingenberg, M. (1996) Mitochondrial ADP/ATP carrier can be reversibly converted into a large channel by Ca^{2+}. *Biochemistry 35*, 8483–8488.

Bushnell, G. W., Louie, G. V., Brayer, G. D. (1990) High-resolution three-dimensional structure of horse cytochrome *c*. *J. Mol. Biol. 214*, 585-595.

Chen, Y.., Liang, Q.., Arciero, D. M.., Hooper, A. B.., Timkovich, R. (2007) Heme crevice disorder after sixth ligand displacement in the cytochrome c-551 family. *Arch. Biochem. Biophys. 457*, 95-104.

Colón, W.., Wakem, L. P.., Sherman, F.., Roder, H. (1997) Identification of the predominant non-native histidine ligand in unfolded cytochrome c. *Biochemistry 36*, 12535-12541.

Dabkowski, E. R., Williamson, C. L., Hollander, J. M. (2008) Mitochondria-specific transgenic overexpression of phospholipid hydroperoxide glutathione peroxidase (GPx4) attenuates ischemia/reperfusion-associated cardiac dysfunction. *Free Radical Bio Med. 45*, 855-865.

Dejean, L. M., Martinez-Caballero, S., Guo, L., Hughes, C., Teijido, O., Ducret, T., Ichas, F., Korsmeyer, S. J., Antonsson, B., Jonas, E. A., Kinnally, K. W. (2005) Oligomeric Bax is a component of the putative cytochrome c release channel MAC, mitochondrial apoptosis-induced channel. *Mol. Biol. Cell 16*, 2424-2432.

Dejean, L. M., Martinez-Caballero, S., Kinnally, K. W. (2006) Is MAC the knife that cuts cytochrome c from mitochondria during apoptosis? *Cell Death Differ. 13*, 1387-1395.

Dickerson, R. E.., Takano, T.., Eisenberg, D.., Kallai, O. B.., Samson, L.., Cooper, A.., Margoliash, E. (1971) Ferricytochrome c. I. General features of the horse and bonito proteins at 2.8 A resolution. *J. Biol. Chem. 246*, 1511-1535.

Eble, K.S., Coleman, W. B., Hantgan, R. R., Cunningham, C. (1990). Tightly associated cardiolipin in the bovine heart mitochondrial ATP synthase as analyzed by 31P nuclear magnetic resonance spectroscopy. *J. Biol. Chem. 265*, 19434-19440.

Elöve, G. A.., Chaffotte, A. F.., Roder, H.., Goldberg, M. E. (1992) Early steps in cytochrome c folding probed by time-resolved circular dichroism and fluorescence spectroscopy. *Biochemistry 31*, 6876-6883.

Elöve, G. A., Bhuyan, A. K., Roder, H. (1994) Kinetic mechanism of cytochrome c folding: involvement of the heme and its ligands. *Biochemistry 33*, 6925-6935.

Epp, O., Ladenstein, R., Wendel, A. (1983) The refined structure of the selenoenzyme glutathione peroxidase at 0.2-nm resolution. *Eur. J. Biochem. 133*, 51-69.

Fisher, W. R., Taniuchi, H., Anfinsen, C. B. (1973) On the role of heme in the formation of the structure of cytochrome c. *J. Biol. Chem. 248*, 3188-3195.

Flohé, L., Loschen, G., Günzler, W. A., Eichele, E. (1972) Glutathione peroxidase, V. The kinetic mechanism. *Hoppe Seylers Z. Physiol. Chem. 353*, 987-999.

Fulda, S., Galluzzi, L., Kroemer, G. (2010) Targeting mitochondria for cancer therapy. *Nat. Rev. Drug Discov. 9*, 447-464.

Gohil, V. M., Greenberg, M. L. (2009) Mitochondrial membrane biogenesis: phospholipids and proteins go hand in hand. *J. Cell Biol. 18*, 469-472.

Gomez, B. Jr., Robinson, N. C. (1999) Phospholipase digestion of bound cardiolipin reversibly inactivates bovine cytochrome bc1. *Biochemistry 38*, 9031-9038.

Hanske, J., Toffey, J. R., Morenz, A. M., Bonilla, A. J., Schiavoni, K. H., Pletneva, E. V. (2012) Conformational properties of cardiolipin-bound cytochrome c. *P. Natl. Acad. Sci. USA 109*, 125-130.

Hauser, H., Pascher, I., Sundell, S. (1988) Preferred conformation and dynamics of the glycerol backbone in phospholipids. An NMR and X-ray single-crystal analysis. *Biochemistry 27*, 9166-9174.

Huang, Z., Jiang, J., Tyurin, V. A., Zhao, Q., Mnuskin, A., Ren, J., Belikova, N. A., Feng, W., Kurnikov, I. V., Kagan, V. E. (2008) Cardiolipin deficiency leads to decreased cardiolipin peroxidation and increased resistance of cells to apoptosis. *Free Radical Bio Med. 44*, 1935-1944.

Imai, H., Suzuki, K., Ishizaka, K., Ichinose, S., Oshima, H., Okayasu, I., Emoto, K., Umeda, M., Nakagawa, Y. (2001) Failure of the expression of phospholipid hydroperoxide glutathione peroxidase in the spermatozoa of human infertile males. *Biol. Reprod. 64*, 674-683.

Imai, H., Koumura, T., Nakajima, R., Nomura, K., Nakagawa, Y. (2003) Protection from inactivation of the adenine nucleotide translocator during hypoglycaemia-induced apoptosis by mitochondrial phospholipid hydroperoxide glutathione peroxidase. *Biochem. J. 371*, 799-809.

Jiang, J., Huang, Z., Zhao, Q., Feng, W., Belikova, N. A., Kagan, V. E. (2008) Interplay between bax, reactive oxygen species production, and cardiolipin oxidation during apoptosis. *Biochem. Biophys. Res. Commun. 368*, 145-150.

Kagan, V. E., Tyurin, V. A., Jiang, J., Tyurina, Y. Y., Ritov, V. B., Amoscato, A. A., Osipov, A. N., Belikova, N. A., Kapralov, A. A., Kini, V., Vlasova, I. I., Zhao, Q., Zou, M., Di, P., Svistunenko, D. A., Kurnikov, I. V., Borisenko, G. G. (2005) Cytochrome c acts as a cardiolipin oxygenase required for release of proapoptotic factors. *Nat. Chem. Biol. 1*, 223-232.

Kapralov, A. A., Yanamala, N., Tyurina, Y. Y., Castro, L., Samhan-Arias, A., Vladimirov, Y. A., Maeda, A., Weitz, A. A., Peterson, J., Mylnikov, D., Demicheli, V., Tortora, V., Klein-Seetharaman, J., Radi, R., Kagan, V. E. (2011) Topography of tyrosine residues and their involvement in peroxidation of polyunsaturated cardiolipin in cytochrome c/cardiolipin peroxidase complexes. *Biochim. Biophys. Acta 1808*, 214721-214755.

Kroemer, G., Reed, J. C. (2000) Mitochondrial control of cell death. *Nat. Med. 6*, 513-519.

Leung, A. W., Halestrap, A. P. (2008) Recent progress in elucidating the molecular mechanism of the mitochondrial permeability transition pore. *Biochim. Biophys. Acta 1777*, 946-952.

Li, H., Zhu, H., Xu, C. J., Yuan, J. (1998) Cleavage of BID by caspase 8 mediates the mitochondrial damage in the Fas pathway of apoptosis. *Cell 94*, 491-501.

Maiorino, M., Ursini, F., Bosello, V., Toppo, S., Tosatto, S. C., Mauri, P., Becker, K., Roveri, A., Bulato, C., Benazzi, L., De Palma, A., Flohé, L. (2007) The thioredoxin specificity of Drosophila GPx: a paradigm for a peroxiredoxin-like mechanism of many glutathione peroxidases. *J. Mol. Biol. 365*, 1033-1046.

Maity, H., Maity, M., Krishna, M. M., Mayne, L., Englander, S. W. (2005) Protein folding: the stepwise assembly of foldon units. *P. Natl. Acad. Sci USA 102*, 4741-4746.

Mannella, C. A. (1992) The 'ins' and 'outs' of mitochondrial membrane channels. *Trends Biochem. Sci 17*, 315-320.

Margoliash, E., Smith, E. L., Kreil, G., Tuppy, H. (1961) Amino-acid sequence of horse heart cytochrome *c*: the complete amino-acid sequence. *Nature 4808*, 1125-1127.

Marí, M., Colell, A., Morales, A., Caballero, F., Moles, A., Fernández, A., Terrones, O., Basañez, G., Antonsson, B., García-Ruiz, C., Fernández-Checa, J. C. (2008) Mechanism of mitochondrial glutathione-dependent hepatocellular susceptibility to TNF despite NF-kappaB activation. *Gastroenterology 134*, 1507-1520.

Muller, F. L., Lustgarten, M. S., Jang, Y., Richardson, A., Van Remmen, H. (2007) Trends in oxidative aging theories. *Free Radical Bio Med. 43*, 477-503.

Nakagawa, T., Shimizu, S., Watanabe, T., Yamaguchi, O., Otsu, K., Yamagata, H., Inohara, H., Kubo, T., Tsujimoto, Y. (2005) Cyclophilin D-dependent mitochondrial permeability transition regulates some necrotic but not apoptotic cell death. *Nature 434*, 652-658.

Nomura, K., Imai, H., Koumura, T., Arai, M., Nakagawa, Y. (1999) Mitochondrial phospholipid hydroperoxide glutathione peroxidase suppresses apoptosis mediated by a mitochondrial death pathway. *J. Biol. Chem. 274*, 29294-29302.

Nomura, K., Imai, H., Koumura, T., Kobayashi, T., Nakagawa, Y. (2000) Mitochondrial phospholipid hydroperoxide glutathione peroxidase inhibits the release of cytochrome c from mitochondria by suppressing the peroxidation of cardiolipin in hypoglycaemia-induced apoptosis. *Biochem. J. 351*, 183-193.

Nury, H., Dahout-Gonzalez, C., Trézéguet, V., Lauquin, G., Brandolin, G., Pebay-Peyroula, E. (2005) Structural basis for lipid-mediated interactions between mitochondrial ADP/ATP carrier monomers. *FEBS Lett. 579*, 6031-6036.

Osman, C., Voelker, D. R., Langer, T. (2011) Making heads or tails of phospholipids in mitochondria. *J. Cell Biol. 192*, 7-16.

Petrosillo, G., Casanova, G., Matera, M., Ruggiero, F. M., Paradies, G. (2006) Interaction of peroxidized cardiolipin with rat-heart mitochondrial membranes: induction of permeability transition and cytochrome c release. *FEBS Lett. 580*, 6311-6316.

Pinheiro, T. J., Elöve, G. A., Watts, A., Roder, H. (1997) Structural and kinetic description of cytochrome c unfolding induced by the interaction with lipid vesicles. *Biochemistry 36*, 13122-13132.

Prats, M., Teissie´, J., Toccane, J. F. (1986) Lateral proton conduction at lipid–water interfaces and its implications for the chemiosmotic-coupling hypothesis. *Nature 322*, 756-758.

Roder, H., Elöve, G. A., Englander, S. W. (1988) Structural characterization of folding intermediates in cytochrome c by H-exchange labelling and proton NMR. *Nature 335*, 700-704.

Russell, B. S., Melenkivitz, R., Bren, K. L. (2000) NMR investigation of ferricytochrome c unfolding: detection of an equilibrium unfolding intermediate and residual structure in the denatured state. *P. Natl. Acad. Sci USA 97*, 8312-8317.

Rytömaa, M., Kinnunen, P. K. (1994) Evidence for two distinct acidic phospholipid-binding sites in cytochrome *c*. *J. Biol. Chem. 269*, 1770-1774.

Rytömaa, M., Kinnunen, P. K. (1995) Reversibility of the binding of cytochrome c to liposomes. Implications for lipid-protein interactions. *J. Biol. Chem. 270*, 3197-3202.

Sanchez, K. M., Schlamadinger, D. E., Gable, J. E., Kim, J. E. (2008) Förster resonance energy transfer and conformational stability of proteins: an advanced biophysical module for physical chemistry students. *Chem. Educ. 85*, 1253-1256.

Schneider, M., Förster, H., Boersma, A., Seiler, A., Wehnes, H., Sinowatz, F., Neumüller, C., Deutsch, M. J., Walch, A., Hrabé de Angelis, M., Wurst, W., Ursini, F., Roveri, A., Maleszewski, M., Maiorino, M., Conrad, M. (2009) Mitochondrial glutathione peroxidase 4 disruption causes male infertility. *FASEB J. 23*, 3233-3242.

Scheerer, P., Borchert, A., Krauss, N., Wessner, H., Gerth, C., Höhne, W., Kuhn, H. (2007) Structural basis for catalytic activity and enzyme polymerization of phospholipid hydroperoxide glutathione peroxidase-4 (GPx4). *Biochemistry 46*, 9041-9049.

Simpkin, .D, Palmer, G., Devlin, F. J., McKenna, M. C., Jensen, G. M., Stephens, P. J. (1989) The axial ligands of heme in cytochromes: a near-infrared magnetic circular dichroism study of yeast cytochromes c, c1, and b and spinach cytochrome f. *Biochemistry_28*, 8033-8039.

Stepanov, G., Gnedenko, O., Mol'nar, A., Ivanov, A., Vladimirov, Y., Osipov, A. (2009) Evaluation of cytochrome c affinity to anionic phospholipids by means of surface plasmon resonance. *FEBS Lett. 583*, 97-100.

Suen, D. F., Norris, K. L., Youle, R. J. (2008) Mitochondrial dynamics and apoptosis. *Gene Dev. 22*, 1577-1590.

Sun, M. G., Williams, J., Munoz-Pinedo, C., Perkins, G. A., Brown, J. M., Ellisman, M. H., Green, D. R., Frey, T. G. (2007) Correlated three-dimensional light and electron microscopy reveals transformation of mitochondria during apoptosis. *Nat. Cell Biol. 9*, 1057-1065.

Sunde, R. A., Hadley, K. B. (2010) Phospholipid hydroperoxide glutathione peroxidase (Gpx4) is highly regulated in male turkey poults and can be used to determine dietary selenium requirements. *Exp. Biol. Med. 235*, 23-31.

Svistunenko, D. A. (2005) Reaction of haem containing proteins and enzymes with hydroperoxides: the radical view. *Biochim. Biophys. Acta. 1707*, 127-155.

Takahashi, S., Yeh, S. R., Das, T. K., Chan, C. K., Gottfried, D. S., Rousseau, D. L. (1997) Folding of cytochrome *c* initiated by submillisecond mixing. *Nat. Struct. Mol. Biol.* ***4***, 44-50.

Tuominen, E. K., Wallace, C. J., Kinnunen, P. K. (2002) Phospholipid-cytochrome c interaction: evidence for the extended lipid anchorage. *J. Biol. Chem. 277*, 8822-8826.

Youle, R. J., Strasser, A. (2008) The BCL-2 protein family: opposing activities that mediate cell death. *Nat. Rev. Mol.* Cell *Biol. 9*, 47-59.

Zoratti, M., Szabò, I., De Marchi, U. (2005) Mitochondrial permeability transitions: how many doors to the house? *Biochim. Biophys. Acta 1706*, 40-52.

Zoratti, M., De Marchi, U., Gulbins, E., Szabò I. (2009) Novel channels of the inner mitochondrial membrane. *Biochim. Biophys. Acta 1787*, 351-363.

In: Peroxidases
Editors: Lucas Bogaert and Noah Coppens
ISBN: 978-1-62808-261-6

Chapter IV

Glutathione Peroxidases in Stem, Cancer, and Cancer Stem Cells

Shanchun Guo[1]*, Lily Yang[2], Guangdi Wang[3] and Alexzander A. A. Asea[1]

[1]Department of Microbiology, Biochemistry & Immunology, Morehouse School of Medicine, Atlanta, GA

[2]Department of Surgery, Emory University School of Medicine, Atlanta, GA

[3]Department of Chemistry, Xavier University of Louisiana, New Orleans, LA

Abstract

Oxidative stress, as a disturbance in the equilibrium between free radicals (FR), reactive oxygen species (ROS), and endogenous antioxidant defense mechanisms is caused by an imbalance between the production of reactive oxygen and a biological system's ability to readily detoxify the reactive intermediates or easily repair the resulting damage. Mounting evidence has implicated that oxidative stress is involved in various physiological and pathological processes, including DNA damage, proliferation, cell adhesion, and survival of cancer cells. Glutathione peroxidases (GPxs) (EC 1.11.1.9), which can reduce lipid hydroperoxides to their corresponding alcohols and free hydrogen peroxide to water, are enzyme family with peroxidase activity whose main biological roles are to protect the organism from oxidative damage. So far 8 sub-members of GPxs have been identified in humans, i.e. GPx 1–8. They all can react with H_2O_2 and soluble fatty acid hydroperoxides. A large number of publications have demonstrated that GPxs have significant roles in different stages of carcinogenesis. In this book chapter, we will review recent study progress of GPxs in stem, cancer and cancer stem cells. This knowledge will shed light on mechanisms that lead to better modes of cancer treatment.

* Corresponding author: Shanchun Guo, E-mail: sguo@msm.edu

1. Introduction

Oxidative stress is defined as a condition in which the equilibrium between free radicals (FR), reactive oxygen species (ROS), and endogenous antioxidant defense mechanisms in the cell is disturbed; cellular molecules and components undergo severe oxidative damage, ultimately compromising cells viability [1-2] .

Oxidative stress has been demonstrated to be involved in various physiological and pathological processes, including DNA damage, proliferation, cell adhesion, and survival. Moreover, experimental and clinical data also provide compelling evidence for the involvement of oxidative stress in large number of pathological states, such as Parkinson's disease [3-4], Alzheimer's disease [5-7], atherosclerosis [8-9], heart failure [10-11], myocardial infarction [12], fragile X syndrome [13], Sickle cell disease [14], hepatic encephalopathy [15-16], as well as carcinogenesis [17-20]. Glutathione peroxidases (GPxs) (EC 1.11.1.9), which can reduce lipid hydroperoxides to their corresponding alcohols and reduce free hydrogen peroxide to water, is an enzyme family with peroxidase activity whose main biological roles are to protect the organism from oxidative stress damage [21-24]. GPxs include five selenium-containing glutathione peroxidases that are found in different cell fractions and tissues of the body. As expected, Gpxs, as a major defender against oxidative stress was also reported to be involved in Parkinson's disease [25-27], Alzheimer's disease [28-30], atherosclerosis [31-32], heart failure [33-34], myocardial infarction [35-36], Sickle cell disease [37], as well as carcinogenesis [24, 38-39]. In this chapter we will focus on the role and function of GPxs in the regulation of stem cells and cancer stem cells (CSC), highlighting the GPxs-mediated signaling pathways and their potential as a drug target and biomarker. Additionally, we will review and discuss recently identified molecular mechanisms of GPxs in CSC, carcinogenesis, as well as its crosstalk with other signaling pathways.

2. Expression and Function of GPxs in Mammalian Cells

GPxs in vertebrates comprise 8 sub-members, i.e. GPx 1–8. The molecular mass of the active purified mammalian GPx1 has been estimated to be between 83 and 95 kDa, consisting of a tetramer of identical subunits of ~22–23 kDa [40-42]. GPx1 is ubiquitously expressed and dominantly found in the cytosol [43] and mitochondria [44]. The different location of GPx1 may have different function, i.e. GPx1 in cytosol may scavenge hydrophilic peroxide species such as H_2O_2 [43], whereas in mitochondria it may protect mitochondrial DNA from oxidative damage [44-45]. GPx1 is considered the major antioxidant enzyme within the GPx family, although GPx1-/- mice apparently were healthy, fertile and showed no increased sensitivity to hyperoxia and could compensate mild oxidative stress [46].

GPx2 is expressed in the gastrointestinal system of mammals and in humans also in the liver [47-48]. Its expression pattern suggests that major function of GPx2 is against ROS derived from the gut. GPx2 knockout mice do not develop an aberrant phenotype before birth, but GPx2 gene is able to compensate for the lack of Gpx1 gene expression in the ileum epithelium, thus GPx2 may have abolished the protective effect in GPx1-KO mice against the radiation damage in the ileum [49].

GPx3 mainly expressed in the proximal tubuli of the kidney is a secreted plasma protein and was found in most extracellular fluids [43, 50]. GPx3-/- mice show no abnormal phenotype throughout their life time and GPx3 does not serve in the regulation of selenium metabolism [51]. The specific binding of a large pool of GPx3 to basement membranes in the kidney cortex strongly suggests that GPx3 may have glutathione peroxidase activity in the cortical peritubular space [51].

Three distinct GPx4 isoforms with different subcellular localizations are produced in mouse and rat: cytosolic GPx4, mitochondrial GPx4 (mGPx4), and nuclear GPx4 (nGPx4) [52]. Cytosolic GPx4 is essential for embryonic development and cell survival, while, mGPx4 and nGPx4 have been implicated in spermatogenesis and male fertility [52]. The key features of GPx4 function are that it has both anti-oxidative and anti-apoptotic activities [53]. In developing embryos GPx4 expression correlates with areas of reduced apoptosis in developing limbs [54]. In contrast to the other GPx isoenzymes, GPx1-3, all GPx4 knockout strategies failed to produce viable homozygous offspring [55-56].

GPx5, the closest homologue to GPx3, is specifically expressed in the epididymis in the mammalian male reproductive tract, and is androgen-regulated [57]. The kinetics and substrate specificities of GPx5 are not fully understood. But a function connected to the maintenance of sperm DNA integrity by removing excess hydroperoxides can be deduced from GPx5 knockout mice [58]. Thus, GPx5 might be a potent antioxidant scavenger that protects spermatozoa from oxidative injuries that could compromise their integrity and, consequently, embryo viability [58].

GPx6, as a putative odorant binding and metabolizing enzyme involved in cellular biotransformation pathways, was identified by in silico analysis [59]. Its expression appears to be restricted to embryos and the Bowman's glands, the site of several previously identified olfactory-specific biotransformation enzymes [60]. Since GPx6 has not been purified and kinetic analyses do not exist, the knowledge on this GPx is very limited, even as far as the olfactory system is concerned.

GPx7, first identified as a novel glutathione peroxidase with a cysteine instead of Sec in the catalytic centre in Brca1-null mouse embryonic fibroblasts, is a cytoplasmic protein with molecular mass of approximately 22 kDa, [61]. GPx7 has little detectable glutathione peroxidase activity *in vitro* [61]. GPx7 has been found to be located in the lumen of the endoplasmic reticulum [62]. Recently, GPx7 was identified as a stress sensor that transmits oxidative stress signals by forming the disulfide bond between its Cys57 and Cys86 residues and it is essential for releasing excessive ER stress by enhancing GRP78 chaperone activity to maintain physiological homeostasis [63].

GPx8 has been identified in a phylogenetic analysis as a novel member belonging to the GPx family in mammalia and amphibia [64]. GPx8 is a membrane protein of the endoplasmic reticulum and belongs to the lung-abundant enzymes [62, 65]. However, little is known about its function.

As described above, the functions of most of these proteins are not completely known, but they may all be capable of reducing hydroperoxides: $ROOH + 2GSH \rightarrow ROH + H_2O + GSSG$. A common function of GPxs, therefore, should be related to the removal and/or metabolism of hydroperoxides.

3. GPxs in the Regulation of Stem Cells

Stem cell (SC) research has received increasing attention in the past decade due to their invaluable potentials in the clinical applications to cure degenerative diseases, genetic disorders and even cancers. SCs differ from other kinds of cells in the body and are found in all multicellular organisms, that can divide and differentiate into diverse specialized cell types and can self-renew to produce more SCs [66]. In mammals, there are two broad types of SCs: embryonic SCs, which are isolated from the inner cell mass of blastocysts, and adult SCs, which are found in various tissues. Although a great number of studies have been conducted in elucidating the molecular mechanisms involved in the regulation of self-renewal and differentiation of all kinds of progenies that can replenish the cell pools, relatively little effort has been made in studying the metabolic aspects of SCs [66-67].

Since both embryonic SCs and adult SCs are the source of different types of mature cells and that the SC persists over long periods of time, SC should require specific protection from the long-term effects of ROS and oxidative damage. The intersection of ROS and SC function was proven from the analysis of mice deficient in the gene ataxia telangiectasia mutated (Atm). Most Atm knockout mice die in the first few months from an aggressive thymic lymphoma, whereas a small percentage of mice survive and are cancer free. Analysis of these surviving mice demonstrated that by 6 months of age, they all develop a progressive pancytopenia with a decline in red cells, white cells, and platelets [68]. Further analysis showed that Atm–/– hematopoietic SCs (HSC) had a profound defect in self-renewal and a marked increase in ROS levels [68]. Several studies have also demonstrated that deficiencies in FoxO activity lead to oxidative stress and subsequent defects in SC function. Analysis of HSCs from conditional deletion of FoxO1, FoxO3, and FoxO4 mice demonstrated a decline in long-term repopulating activity and an increase in ROS levels [69]. Consistent with its role in regulating antioxidant defenses, FoxO-deficient SCs exhibited reduced levels of MnSOD and catalase [69]. In a recent study, Prdm16 deficiency was proven to lead to changes in the levels of ROS, depletion of SCs, increased cell death and altered cell-cycle distribution in the haematopoietic and nervous systems [70]. Prdm16 is a transcription factor that regulates leukaemogenesis, palatogenesis and brown-fat development [71-73]. In neural SCs, Prdm16 binds to the Hgf promoter, and Hgf expression declined in the absence of Prdm16. Administration of the anti-oxidant, N-acetyl-cysteine, to Prdm16-deficient mice partially rescued defects in neural stem/progenitor cell function and neural development [70]. Prdm16 therefore promotes SC maintenance in multiple tissues, partly by modulating oxidative stress. All these results involving Atm, FoxO, as well as Prdm16 indicate that there is a strong connection between ROS homeostasis and the maintenance of SC function.

With the increasing acknowledgement that hydroperoxides are not simply harmful ROS but mediators of physiological processes including SC regulation, the role of GPxs in the regulation of SCs is finally being recognized. GPx1 is one of the most abundant members of the GPx family of enzymes that include an epithelial-specific enzyme highly expressed in intestine, GPx2; a secreted subtype, GPx3 and GPx4, which are widely expressed and differ in their substrate specificity compared to the other family members. Accordingly, GPx1 is a crucial antioxidant enzyme involved in preventing the harmful accumulation of intracellular hydrogen peroxide. Although GPx1-/- mice apparently were healthy [46], GPx1 was still considered the most important member of GPx family in modulating cellular oxidant stress and redox-mediated responses, as well as in the regulation of SCs [21, 74]. Localization of GPx2, the gastrointestinal form of GPx's, in the intestinal crypt epithelium points to a specific function of this particular GPx in gastrointestinal SCl regulation. Loss of GPx2 led to an increase in apoptotic cells at colonic crypt bases, an area essential for the self-renewal of the intestinal epithelium. Moreover, mitotic cells increased in the middle parts of the crypts, indicating an extension of the proliferative area [75]. These findings indicate a role for GPx2 in regulating mucosal SC homeostasis. Recently, GPx3 was demonstrated to be a new retinoid target gene and crucial for human skeletal muscle precursor cell survival [76]. In an early report, in normal HSCs engineered to express Gpx3 short hairpin RNA (shRNA) were much less competitive in vivo than control cells. Consistent with this, HSCs overexpressing Gpx3 were significantly more competitive than control cells in long-term repopulation experiments, and overexpression of the self-renewal genes Prdm16 or Hoxb4 boosted Gpx3 expression [77]. As mentioned above, GPx4 is essential for embryonic SC development, survival, especially in spermatogenesis and male fertility [52] and GPx5 may also have a role in the mammalian male SC regulation [57]. So far, there has been no report about the role of GPx6, 7 and 8 in SC regulation.

4. GPxs in Several Types of Tumor

Many studies have reported changes in GPx levels in several types of tumor. However, it is difficult to know which of these differences are causative factors in the progression of the cancers and which are due to the acquisition of the transformed state that is typically associated with deregulation of a large number of genes.

The role of GPx1 in tumor has been recently reviewed [21, 60]. Since GPx1 was reported to prevent oxidative DNA mutations and counteracts the production of COX/LOX-derived proinflammatory mediators such as prostaglandins and leukotrienes, GPx1 may prevent tumorigenesis [78]. GPx1 overexpression reduced growth of tumor cell lines in vitro and in a xenograft model indicates a role of protective effect of GPx1 in tumorigenesis [79]. The dual role of GPx2 in tumorigenesis has been discussed in recent reviews [38, 60]. Overexpression of GPx2 was observed in several tumors including colorectal cancer [80-81], Barrett's esophagus squamous cell carcinoma [82-83] and lung adenocarcinomas [84], indicating that GPx2 may be an oncogene. However, GPx2 was also observed to be down-regulated in prostatic intraepithelial neoplasia [84], indicating GPx2 may play a more complex role in tumorigenesis. GPx3 is considered to be a novel tumor suppressor, since hypermethylation of the GPx3 promoter was detected in tumor samples from patients with Barrett's esophagus [85-87], endometrial [88] and prostate cancer [89] and down-regulation was generally associated with a poor prognosis. GPx4 is also considered to be a tumor suppressor, since it was down-regulated in pancreatic [79] and breast cancer [90]. In addition, GPx4 overexpression reduced tumor growth of tumorigenic L929 fibrosarcoma cells [91]. So far, there is no report about the role of GPx5, 6, 7 and 8 in tumorigenesis.

To further define the roles of GPxs and clinical significance in cancer, we used an online database [92-93] to assess relevance of GPxs expression to relapse free survival in breast and lung cancer patients. The database was established using gene expression data and survival information of 2,878 breast cancer patients and 1,404 lung cancer patients downloaded from

Gene Expression Omnibus (GEO) (Affymetrix HGU133A and HGU133+2 microarrays). Briefly, GPx genes were entered into the database (http://kmplot.com/breast/) to obtain Kaplan-Meier survival plots where the number-at-risk is indicated below the main plot. Hazard ratio (and 95% confidence intervals) and log rank P were calculated and displayed on the webpage. Kaplan-Meier survival plots demonstrate the prognostic relevance of GPx expression on breast or lung cancer patient survival using various data cohorts (Figures1 &2). These correlation data indicates that GPx2, GPx5 may be tumor suppressors in human breast cancer, while GPx3 is a tumor suppressor in human lung cancer, GPx2, GPx6/7 behave as oncogenes in human lung cancer.

5. GPxs in Cancer Stem Cells

Cancer cells are believed to originate from a small subset of malignant cells that have a high capacity of self-renewal and differentiation, namely cancer stem cells (CSCs) or tumor-initiating cells [94-95]. Although their existence and role remain a topic of intense debate, recently, the identification of CSCs in endogenously developing mouse tumors has provided further support for this concept [96-97]. CSCs and SCs share important functional similarities such as 'self-renewal' and differentiation potential. However, CSCs and SCs have different biological behaviours, mainly because of a profound deregulation of self-renewal capability in CSCs. Unlike SCs which have the process of differentiation leading specialized progenies with no proliferative potential, CSCs give rise to progenies that do not undergo terminal differentiation but instead exhibit uncontrolled proliferation. Differences in mode of division, cell-cycle properties, replicative potential and handling of DNA damage, in addition to the activation/inactivation of cancer-specific molecular pathways confer on cancer stem cells a malignant phenotype. CSCs initiate and drive carcinogenesis and differentiation contributing to tumor cellular heterogeneity through deregulation of the self-renewal process [98]. Thus, the population of CSCs may be a risk factor for carcinogenesis [99]. Many human cancers possess an enhanced tumor-initiating capacity, can undergo self-renewal, partially recreate the cellular heterogeneity of the parental tumor. Those cancer cells seem to be generally more resistant than other cancer cells to conventional anticancer therapeutics. Previous studies have identified CSCs in solid tumors that are responsible for tumor initiation, growth and metastasis [100-101].

In comparison with SCs or cancer cells as a whole population, relatively little is known about the ROS levels in CSCs. Similar to SCs, such as HSCs, CSCs also show lower intracellular ROS contents than non-CSCs, which may be due to the increased expression of free radical scavenging systems [102-103]. In addition, the unchecked ROS accumulation is thought to play a part in the conversion from normal hemopoietic stem cells to leukemic cells [104-105]. In regard to the biological properties of CSCs, this unique cell subpopulation might have a high antioxidant capacity to keep cellular ROS at a moderate level and maintain both stemness and cancer-forming capabilities. CD44+/CD24- breast CSCs have been confirmed to exhibit an enhanced ROS defense system by over-expressing antioxidant enzyme genes and less DNA damage when compared with non-CSCs [102]. Lower levels of basal and radiation-induced ROS in CD44+/CD24- breast CSCs have been observed to relate with tumorigenicity and resistance to radiation [102].

GPx1

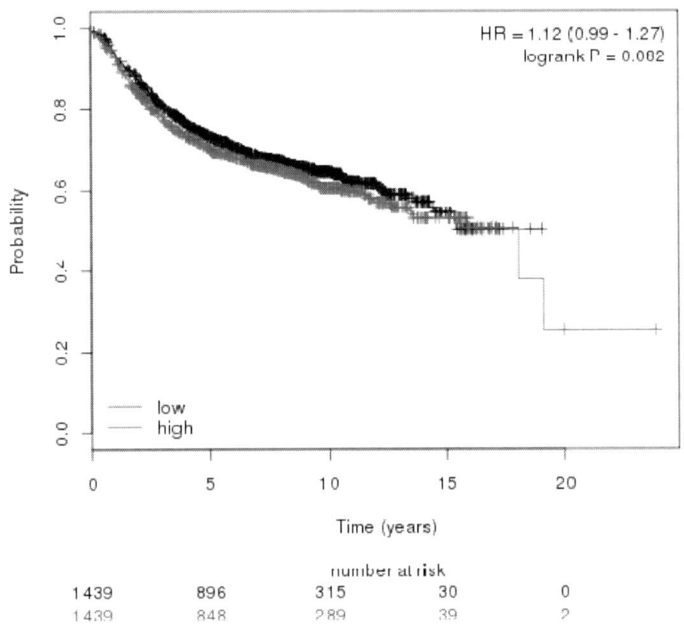
HR = 1.12 (0.99 - 1.27)
logrank P = 0.002
low
high
Probability
0.0
0.2
0.4
0.6
0.8
1.0
0
5
10
15
20
Time (years)
number at risk
1439 896 315 30 0
1439 848 289 39 2

GPx2

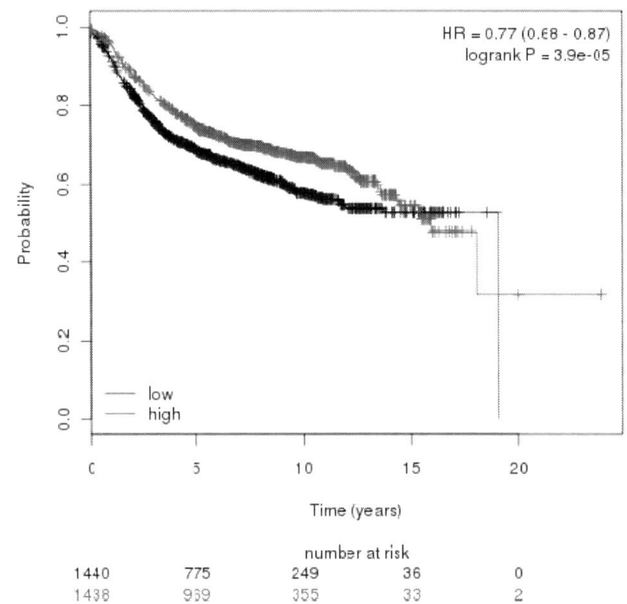
HR = 0.77 (0.68 - 0.87)
logrank P = 3.9e-05
low
high
Probability
0.0
0.2
0.4
0.6
0.8
1.0
0
5
10
15
20
Time (years)
number at risk
1440 775 249 36 0
1438 939 355 33 2

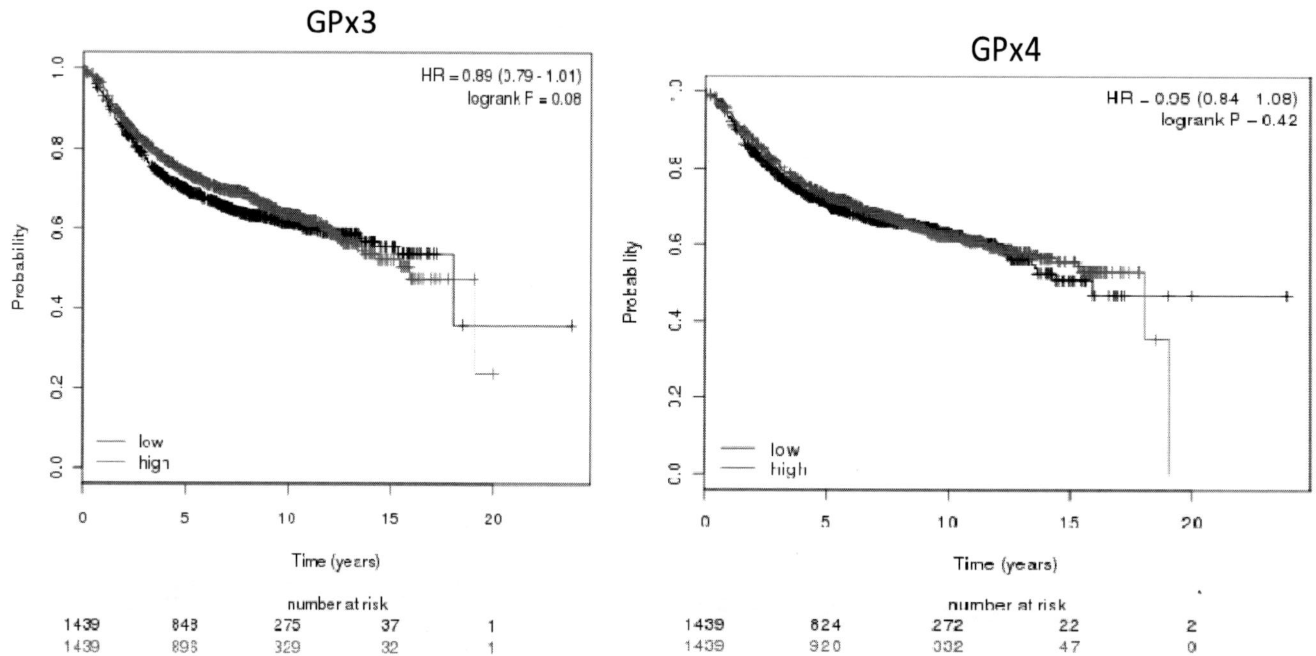

Figure 1. (Continued).

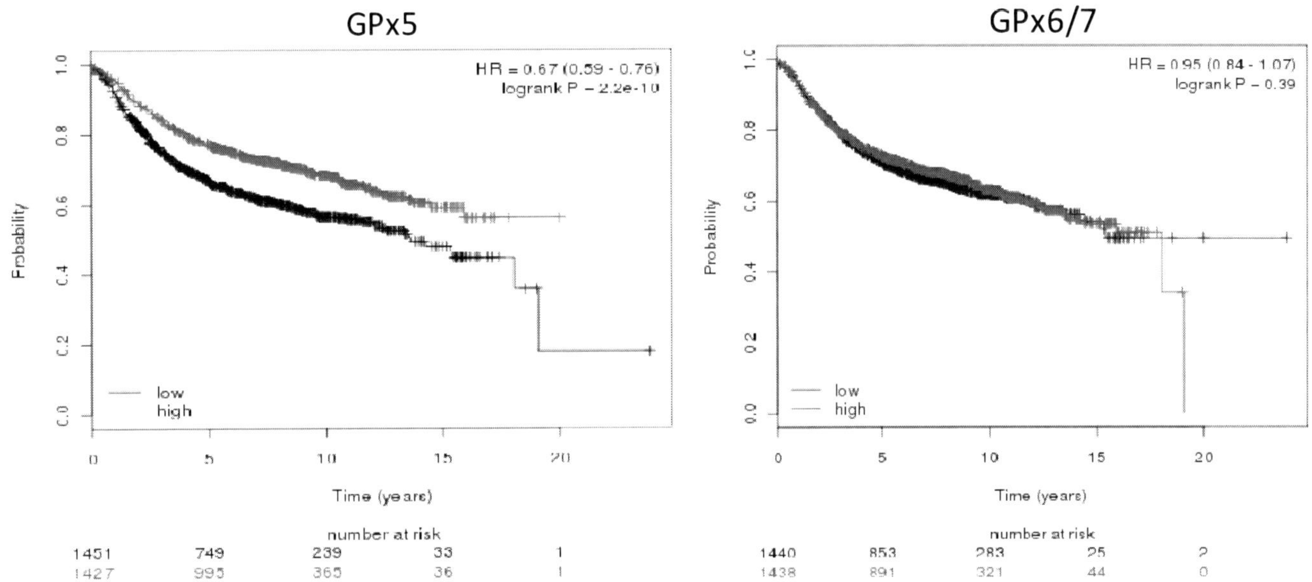

Figure 1. Relevance of GPx mRNA expression to relapse free survival in breast cancer patients. The database was established using gene expression data and survival information of 2,878 breast cancer patients downloaded from Gene Expression Omnibus (GEO) (Affymetrix HGU133A and HGU133+2 microarrays). Kaplan-Meier survival plots demonstrate the prognostic relevance of GPx expression on breast cancer patient survival using various data cohorts.

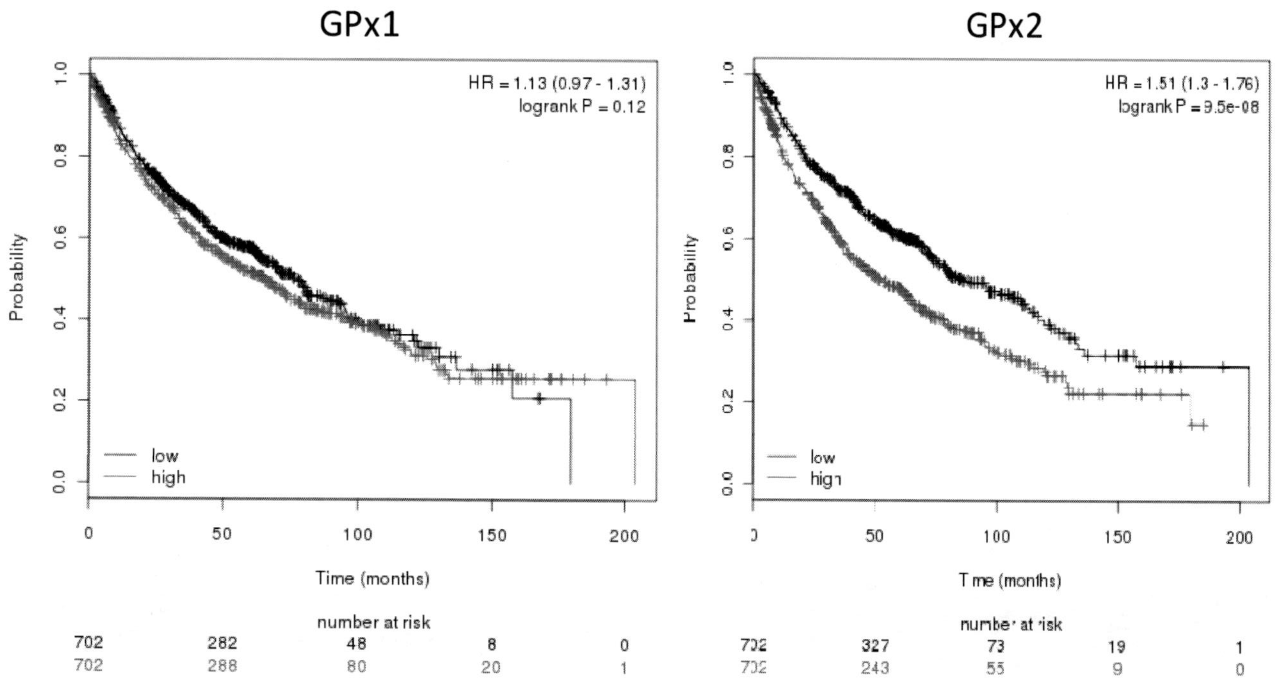

Figure 2. (Continued).

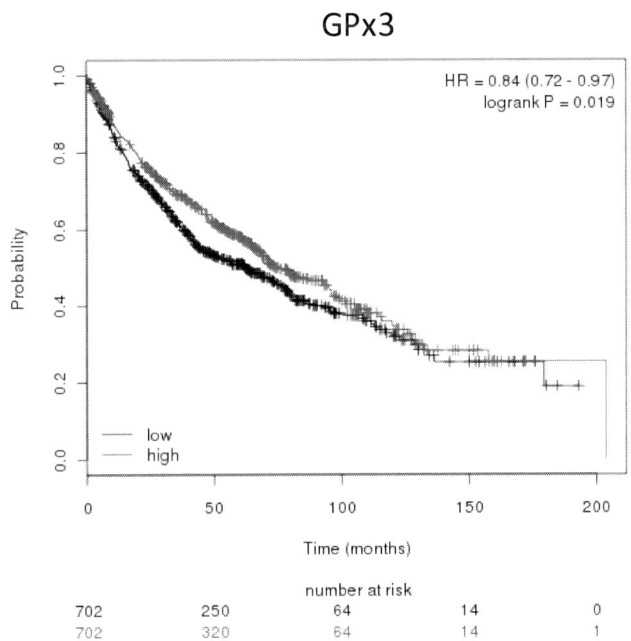
GPx3
HR = 0.84 (0.72 - 0.97)
logrank P = 0.019
Probability
0.0
0.2
0.4
0.6
0.8
1.0
low
high
0
50
100
150
200
Time (months)
number at risk
702 250 64 14 0
702 320 64 14 1

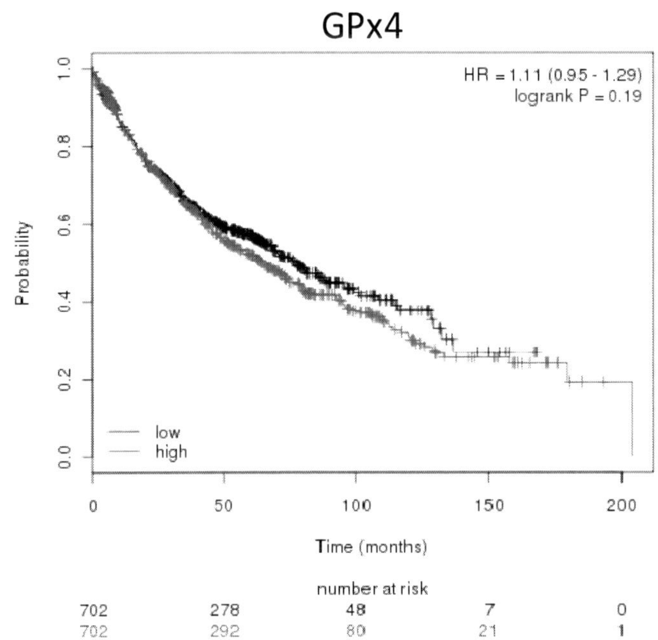
GPx4
HR = 1.11 (0.95 - 1.29)
logrank P = 0.19
Probability
0.0
0.2
0.4
0.6
0.8
1.0
low
high
0
50
100
150
200
Time (months)
number at risk
702 278 48 7 0
702 292 80 21 1

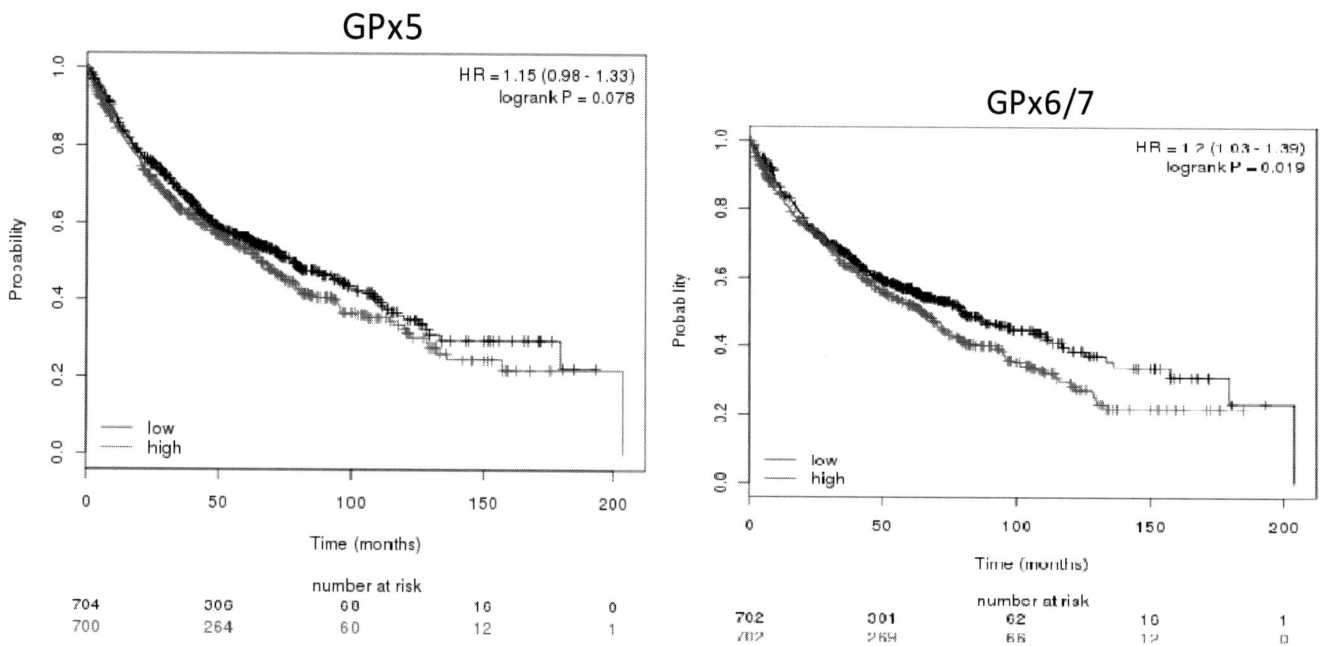

Figure 2. Relevance of GPx mRNA expression to relapse free survival in lung cancer patients.
The database was established using gene expression data and survival information of 1,404 lung cancer patients downloaded from Gene Expression Omnibus (GEO) (Affymetrix HGU133A and HGU133+2 microarrays). Kaplan-Meier survival plots demonstrate the prognostic relevance of GPx expression on lung cancer patient survival using various data cohorts.

Recently, CD13/Aminopeptidase N, which is a scavenger enzyme in the ROS metabolic pathway [106-107] was reported to play a role in supporting the survival of CSCs and that there is an EMT-associated reduction in ROS elevation [108]. CSCs display an EMT phenotype and are resistant to current therapies. More recently, Dong and colleagues reported that these phenotypes in basal-like breast cancer are promoted by a metabolic switch to glucose metabolism, resulting in decreased ROS levels [109]. They observed that loss of fructose-1,6-biphosphatase (FBP1) induces glycolysis and results in increased glucose uptake, macromolecule biosynthesis, formation of tetrameric PKM2, and maintenance of ATP production under hypoxia. Loss of FBP1 also inhibits oxygen consumption and ROS production by suppressing mitochondrial complex I activity; this metabolic reprogramming results in an increased CSC-like property and tumorigenicity by enhancing the interaction of β-catenin with T-cell factor.

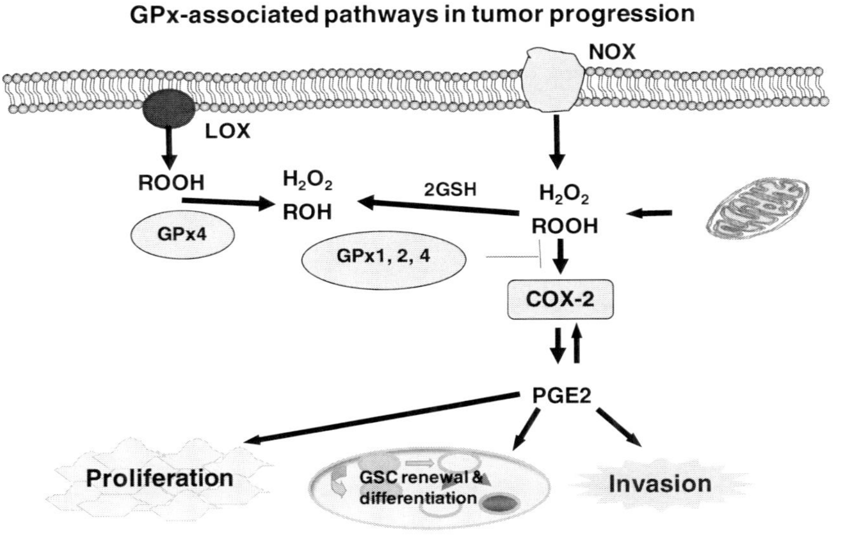

Figure 3. GPxs and COX/LOX activity. Mitochondria and NADPH oxidases (NOX) are intracellular sources of H_2O_2 or other hydroperoxides (ROOH). H_2O_2 and ROOH are reduced by all GPx1, 2, and 4 and also by GPx3. Hydroperoxides activate COX-2 which, in principle, is inhibited by all GPxs, preferentially, however, by GPx4. COX-2 forms PGE2 which, in an autocrine loop can induce the expression of COX-2 to further increase PGE2 production, which promote cancer cell proliferation, invasion and self-renewal and differentiation of CSCs.

NADPH oxidases (NOX) and mitochondria are intracellular sources of H_2O_2 or other hydroperoxides (ROOH) subsequently produced. A number of studies have established that H_2O_2 and/or ROOH are able to activate cyclooxygenase-2 (COX-2), one of the cyclooxygenases which catalyze a key step in the formation of proinflammatory prostaglandins, e.g., PGE2 [24, 110-111]. COX-2 is up-regulated in inflammation and cancer tissue [112-113] and several kinds of CSCs [114-116]. The main product of the COX-2 cascade is PGE2 that is the key mediator in the acute inflammatory response [117], but it is also directly involved in tumor cell proliferation, migration and invasion [118]. PGE2 enhances tumor cell proliferation and inhibits apoptosis by activation of pro-survival pathways such as the PI3K/Akt or the Ras-MAPK/ERK pathways. It supports cancer cell migration and invasion [117-118] and angiogenesis [119-120]. Moreover, PGE2 induces CSCs through the Wnt pathway, a key event especially in carcinogenesis [121-122].

An imbalance or dysregulation of ROS levels may generate cells with abnormal growth and is therefore potentially tumorigenic. Thus, a precise balance between processes generating ROS and those decomposing ROS is critical for tumor development including CSC self-renewal and differentiation. One tier of the cellular protective system against ROS constitutes the GPx family. Due to the significant role GPxs play in modulating cellular oxidant stress and redox-mediated responses, certain members of GPx should also play a critical role in CSC self-renewal and differentiation [60, 123-124]. Glioma stem cell lines expressing active GPx1 might decrease ROS level in Glioma CSCs and resist ROS/RNS-mediated cell death, thus creates a carcinoma stem cell niche [125]. Herault et al also reported that expression of the ROS scavenger, GPx3 positively correlates with the frequency of leukemia stem cells (LSCs) in Hoxa9+Meis1-induced leukemias. Compared to leukemia with a low frequency of LSCs, a leukemia with a high frequency of LSCs showed hypomethylation of the Gpx3 promoter region, and expressed high levels of Gpx3 and low levels of ROS [77]. In addition, GPxs could also directly decrease H_2O_2 and/or ROOH level, then inactivate COX-2 and PEG2 and finally modulate CSCs (Figure 3). Recently, Dr. Weinberg's group show that tumor cells are able to elicit a strong induction of the COX-2/microsomal prostaglandin-E synthase-1 (mPGES-1)/PGE2 axis in MSCs recruited to the tumor-associated stroma by releasing IL-1, which in turn elicits a mesenchymal/stem cell–like phenotype in the carcinoma cells [126]. GPxs, such as GPx1 also significantly impact proinflammatory cytokine-induced redox signaling and human endothelial cell activation [127-129], while endothelial cell activation is considered to be able to create a niche that helps promote self-renewal of CSCs [130-131]. Therefore, GPxs could modulate CSCs through multiple pathways.

6. Conclusion

Much has been learned about the GPx family in redox biology and cancer biology since their discovery as crucial antioxidant enzymes that inactivate peroxides. GPxs are involved in balancing the H_2O_2 homeostasis in signaling cascades and in tumorigenesis. GPx1, GPx2, GPx3 and 4 are also implicated in the self-renewal and differentiation stem and CSC cells through multiple pathways.

The most challenging problem is to unravel how the different GPx members can exert highly specific biological functions. The functions of some members of GPx, such as GPx6, 7 and 8 are still not known and need to be further clarified.

References

[1] Sinha, K., Das, J.; Pal, P. B. & Sil, P. C. (2013). Oxidative stress: the mitochondria-dependent and mitochondria-independent pathways of apoptosis, *Arch Toxicol*, *30*, 30.

[2] Devasagayam, T. P., Tilak, J. C., Boloor, K. K., Sane, K. S., Ghaskadbi, S. S. & Lele, R. D. (2004). Free radicals and antioxidants in human health: current status and future prospects, *J Assoc Physicians India*, *52*, 794-804.

[3] Ciccone, S., Maiani, E., Bellusci, G., Diederich, M. & Gonfloni, S. (2013). Parkinson's disease: a complex interplay of mitochondrial DNA alterations and oxidative stress, *Int J Mol Sci*, *14*, 2388-2409.

[4] Janda, E., Isidoro, C., Carresi, C. & Mollace, V. (2012). Defective autophagy in Parkinson's disease: role of oxidative stress, *Mol Neurobiol*, *46*, 639-661.

[5] Greenough, M. A., Camakaris, J. & Bush, A. I. (2013). Metal dyshomeostasis and oxidative stress in Alzheimer's disease, *Neurochem Int*, *62*, 540-555.

[6] Eskici, G. & Axelsen, P. H. (2012). Copper and oxidative stress in the pathogenesis of Alzheimer's disease, *Biochemistry*, *51*, 6289-6311.

[7] Massaad, C. A. (2011). Neuronal and vascular oxidative stress in Alzheimer's disease, *Curr Neuropharmacol*, *9*, 662-673.

[8] Peluso, I., Morabito, G., Urban, L., Ioannone, F. & Serafini, M. (2012). Oxidative stress in atherosclerosis development: the central role of LDL and oxidative burst, *Endocr Metab Immune Disord Drug Targets*, *12*, 351-360.

[9] Vogiatzi, G., Tousoulis, D. & Stefanadis, C. (2009). The role of oxidative stress in atherosclerosis, *Hellenic J Cardiol*, *50*, 402-409.

[10] Ahmed, Z. & Tang, W. H. (2012). Pharmacologic strategies to target oxidative stress in heart failure, *Curr Heart Fail Rep*, *9*, 14-22.

[11] Tsutsui, H., Kinugawa, S. & Matsushima, S. (2008). Oxidative stress and mitochondrial DNA damage in heart failure, *Circ J*, *72* Suppl A, A31-37.

[12] Di Filippo, C., Cuzzocrea, S., Rossi, F., Marfella, R. & D'Amico, M. (2006). Oxidative stress as the leading cause of acute myocardial infarction in diabetics, *Cardiovasc Drug Rev*, *24*, 77-87.

[13] el Bekay, R., Romero-Zerbo, Y., Decara, J., Sanchez-Salido, L., Del Arco-Herrera, I., Rodriguez-de Fonseca, F. & de Diego-Otero, Y. (2007). Enhanced markers of oxidative stress, altered antioxidants and NADPH-oxidase activation in brains from Fragile X mental retardation 1-deficient mice, a pathological model for Fragile X syndrome, *Eur J Neurosci*, *26*, 3169-3180.

[14] Nur, E., Biemond, B. J., Otten, H. M., Brandjes, D. P. & Schnog, J. J. (2011). Oxidative stress in sickle cell disease; pathophysiology and potential implications for disease management, *Am J Hematol*, *86*, 484-489.

[15] Seyan, A. S., Hughes, R. D. & Shawcross, D. L. (2010). Changing face of hepatic encephalopathy: role of inflammation and oxidative stress, *World J Gastroenterol*, *16*, 3347-3357.

[16] Schliess, F., Gorg, B. & Haussinger, D. (2006). Pathogenetic interplay between osmotic and oxidative stress: the hepatic encephalopathy paradigm, *Biol Chem*, *387*, 1363-1370.

[17] Sohal, R. S. & Weindruch, R. (1996). Oxidative stress, caloric restriction, and aging, *Science*, *273*, 59-63.

[18] Wang, L., Chen, F., Zhang, Z., Chen, G., Luo, J. & Shi, X. (2012). Cancer stem cells in the mechanism of metal carcinogenesis, *J Environ Pathol Toxicol Oncol*, *31*, 245-263.

[19] Ziech, D., Franco, R., Pappa, A. & Panayiotidis, M. I. (2011). Reactive oxygen species (ROS)--induced genetic and epigenetic alterations in human carcinogenesis, *Mutat Res*, *711*, 167-173.

[20] Kryston, T. B., Georgiev, A. B., Pissis, P. & Georgakilas, A. G. (2011). Role of oxidative stress and DNA damage in human carcinogenesis, *Mutat Res*, *711*, 193-201.

[21] Lubos, E., Loscalzo, J. & Handy, D. E. (2011). Glutathione peroxidase-1 in health and disease: from molecular mechanisms to therapeutic opportunities, *Antioxid Redox Signal*, *15*, 1957-1997.

[22] Battin, E. E. & Brumaghim, J. L. (2009). Antioxidant activity of sulfur and selenium: a review of reactive oxygen species scavenging, glutathione peroxidase, and metal-binding antioxidant mechanisms, *Cell Biochem Biophys*, *55*, 1-23.

[23] Toppo, S., Flohe, L., Ursini, F., Vanin, S. & Maiorino, M. (2009). Catalytic mechanisms and specificities of glutathione peroxidases: variations of a basic scheme, *Biochim Biophys Acta*, *1790*, 1486-1500.

[24] Brigelius-Flohe, R. & Kipp, A. Glutathione peroxidases in different stages of carcinogenesis, *Biochim Biophys Acta*, *1790*, 1555-1568.

[25] Kish, S. J., Morito, C. & Hornykiewicz, O. (1985). Glutathione peroxidase activity in Parkinson's disease brain, *Neurosci Lett*, *58*, 343-346.

[26] Johannsen, P., Velander, G., Mai, J., Thorling, E. B. & Dupont, E. (1991). Glutathione peroxidase in early and advanced Parkinson's disease, *J Neurol Neurosurg Psychiatry*, *54*, 679-682.

[27] Thiruchelvam, M., Prokopenko, O., Cory-Slechta, D. A., Buckley, B. & Mirochnitchenko, O. (2005). Overexpression of superoxide dismutase or glutathione peroxidase protects against the paraquat + maneb-induced Parkinson disease phenotype, *J Biol Chem*, *280*, 22530-22539.

[28] Yoo, M. H., Gu, X., Xu, X. M., Kim, J. Y., Carlson, B. A., Patterson, A. D., Cai, H., Gladyshev, V. N. & Hatfield, D. L. (2010). Delineating the role of glutathione peroxidase 4 in protecting cells against lipid hydroperoxide damage and in Alzheimer's disease, *Antioxid Redox Signal*, *12*, 819-827.

[29] Ceballos-Picot, I., Merad-Boudia, M., Nicole, A., Thevenin, M., Hellier, G., Legrain, S. & Berr, C. (1996). Peripheral antioxidant enzyme activities and selenium in elderly subjects and in dementia of Alzheimer's type--place of the extracellular glutathione peroxidase, *Free Radic Biol Med*, *20*, 579-587.

[30] Paz-y-Mino, C., Carrera, C., Lopez-Cortes, A., Munoz, M. J., Cumbal, N., Castro, B., Cabrera, A. & Sanchez, M. E. (2010). Genetic polymorphisms in apolipoprotein E and glutathione peroxidase 1 genes in the Ecuadorian population affected with Alzheimer's disease, *Am J Med Sci*, *340*, 373-377.

[31] Lewis, P., Stefanovic, N., Pete, J., Calkin, A. C., Giunti, S., Thallas-Bonke, V., Jandeleit-Dahm, K. A., Allen, T. J., Kola, I., Cooper, M. E. & de Haan, J. B. (2007). Lack of the antioxidant enzyme glutathione peroxidase-1 accelerates atherosclerosis in diabetic apolipoprotein E-deficient mice, *Circulation*, *115*, 2178-2187.

[32] Lin, M. S., Hsu, H. C., Lin, L. C., Li, H. Y., Lee, B. C., Lee, Y. T. & Chen, M. F. (2007). Higher glutathione peroxidase expression in thoracic aorta as a protective factor against oxidative stress and atherosclerosis in rabbits, *Cardiology*, *108*, 381-386.

[33] Shiomi, T., Tsutsui, H., Matsusaka, H., Murakami, K., Hayashidani, S., Ikeuchi, M., Wen, J., Kubota, T., Utsumi, H. & Takeshita, A. (2004). Overexpression of glutathione peroxidase prevents left ventricular remodeling and failure after myocardial infarction in mice, *Circulation*, *109*, 544-549.

[34] Dabkowski, E. R., Williamson, C. L. & Hollander, J. M. (2008). Mitochondria-specific transgenic overexpression of phospholipid hydroperoxide glutathione peroxidase (GPx4) attenuates ischemia/ reperfusion-associated cardiac dysfunction, *Free Radic Biol Med*, *45*, 855-865.

[35] Yoshida, T., Watanabe, M., Engelman, D. T., Engelman, R. M., Schley, J. A., Maulik, N., Ho, Y. S., Oberley, T. D. & Das, D. K. (1996). Transgenic mice overexpressing glutathione peroxidase are resistant to myocardial ischemia reperfusion injury, *J Mol Cell Cardiol*, *28*, 1759-1767.

[36] Cheng, M. L., Chen, C. M., Ho, H. Y., Li, J. M. & Chiu, D. T. (2009). Effect of acute myocardial infarction on erythrocytic glutathione peroxidase 1 activity and plasma vitamin e levels, *Am J Cardiol*, *103*, 471-475.

[37] Natta, C. L., Chen, L. C. & Chow, C. K. (1990). Selenium and glutathione peroxidase levels in sickle cell anemia, *Acta Haematol*, *83*, 130-132.

[38] Brigelius-Flohe, R. & Kipp, A. P. (2012). Physiological functions of GPx2 and its role in inflammation-triggered carcinogenesis, *Ann N Y Acad Sci*, *1259*, 19-25.

[39] Zhuo, P. & Diamond, A. M. (2009). Molecular mechanisms by which selenoproteins affect cancer risk and progression, *Biochim Biophys Acta*, *1790*, 1546-1554.

[40] Awasthi, Y. C., Beutler, E. & Srivastava, S. K. (1975). Purification and properties of human erythrocyte glutathione peroxidase, *J Biol Chem*, *250*, 5144-5149.

[41] Miwa, T., Adachi, T., Ito, Y., Hirano, K. & Sugiura, M. (1983). Purification and properties of glutathione peroxidase from human liver, *Chem Pharm Bull* (Tokyo), *31*, 179-185.

[42] Oh, S. H., Ganther, H. E. & Hoekstra, W. G. (1974). Selenium as a component of glutathione periodase isolated from ovine erythrocytes, *Biochemistry*, *13*, 1825-1829.

[43] Brigelius-Flohe, R. (1999). Tissue-specific functions of individual glutathione peroxidases, *Free Radic Biol Med*, *27*, 951-965.

[44] Legault, J., Carrier, C., Petrov, P., Renard, P., Remacle, J. & Mirault, M. E. (2000). Mitochondrial GPx1 decreases induced but not basal oxidative damage to mtDNA in T47D cells, *Biochem Biophys Res Commun*, *272*, 416-422.

[45] Kienhofer, J., Haussler, D. J., Ruckelshausen, F., Muessig, E., Weber, K., Pimentel, D., Ullrich, V., Burkle, A. & Bachschmid, M. M. (2009). Association of mitochondrial antioxidant enzymes with mitochondrial DNA as integral nucleoid constituents, *FASEB J*, *23*, 2034-2044.

[46] Ho, Y. S., Magnenat, J. L., Bronson, R. T., Cao, J., Gargano, M., Sugawara, M., Funk, C. D. (1997). Mice deficient in cellular glutathione peroxidase develop normally and show no increased sensitivity to hyperoxia, *J Biol Chem*, *272*, 16644-16651.

[47] Chu, F. F., Doroshow, J. H. & Esworthy, R. S. (1993). Expression, characterization, and tissue distribution of a new cellular selenium-dependent glutathione peroxidase, GSHPx-GI, *J Biol Chem*, *268*, 2571-2576.

[48] Dreher, I., Schmutzler, C., Jakob, F. & Kohrle, J. (1997). Expression of selenoproteins in various rat and human tissues and cell lines, *J Trace Elem Med Biol*, *11*, 83-91.

[49] Esworthy, R. S., Mann, J. R., Sam, M. & Chu, F. F. (2000). Low glutathione peroxidase activity in Gpx1 knockout mice protects jejunum crypts from gamma-irradiation damage, *Am J Physiol Gastrointest Liver Physiol*, *279*, G426-436.

[50] Avissar, N., Ornt, D. B., Yagil, Y., Horowitz, S., Watkins, R. H., Kerl, E. A., Takahashi, K., Palmer, I. S. & Cohen, H. J. (1994). Human kidney proximal tubules are the main source of plasma glutathione peroxidase, *Am J Physiol*, *266*, C367-375.

[51] Olson, G. E., Whitin, J. C., Hill, K. E., Winfrey, V. P., Motley, A. K., Austin, L. M., Deal, J., Cohen, H. J. & Burk, R. F. (2010). Extracellular glutathione peroxidase (Gpx3) binds specifically to basement membranes of mouse renal cortex tubule cells, *Am J Physiol Renal Physiol*, *298*, F1244-1253.

[52] Schneider, M., Forster, H., Boersma, A., Seiler, A., Wehnes, H., Sinowatz, F., Neumuller, C., Deutsch, M. J., Walch, A., Hrabe de Angelis, M., Wurst, W., Ursini, F., Roveri, A., Maleszewski, M., Maiorino, M. & Conrad, M. (2009). Mitochondrial glutathione peroxidase 4 disruption causes male infertility, *FASEB J*, *23*, 3233-3242.

[53] Savaskan, N. E., Ufer, C., Kuhn, H. & Borchert, A. (2007). Molecular biology of glutathione peroxidase 4: from genomic structure to developmental expression and neural function, *Biol Chem*, *388*, 1007-1017.

[54] Schnabel, D., Salas-Vidal, E., Narvaez, V., Sanchez-Carbente Mdel, R., Hernandez-Garcia, D., Cuervo, R. & Covarrubias, L. (2006). Expression and regulation of antioxidant enzymes in the developing limb support a function of ROS in interdigital cell death, *Dev Biol*, *291*, 291-299.

[55] Imai, H., Hirao, F., Sakamoto, T., Sekine, K., Mizukura, Y., Saito, M., Kitamoto, T., Hayasaka, M., Hanaoka, K. & Nakagawa, Y. (2003). Early embryonic lethality caused by targeted disruption of the mouse PHGPx gene, *Biochem Biophys Res Commun*, *305*, 278-286.

[56] Yant, L. J., Ran, Q., Rao, L., Van Remmen, H., Shibatani, T., Belter, J. G., Motta, L., Richardson, A. & Prolla, T. A. (2003). The selenoprotein GPX4 is essential for mouse development and protects from radiation and oxidative damage insults, *Free Radic Biol Med*, *34*, 496-502.

[57] Rejraji, H., Vernet, P. & Drevet, J. R. (2002). GPX5 is present in the mouse caput and cauda epididymidis lumen at three different locations, *Mol Reprod Dev*, *63*, 96-103.

[58] Chabory, E., Damon, C., Lenoir, A., Kauselmann, G., Kern, H., Zevnik, B., Garrel, C., Saez, F., Cadet, R., Henry-Berger, J., Schoor, M., Gottwald, U., Habenicht, U., Drevet, J. R. & Vernet, P. (2009). Epididymis seleno-independent glutathione peroxidase 5 maintains sperm DNA integrity in mice, *J Clin Invest*, *119*, 2074-2085.

[59] Dear, T. N., Campbell, K. & Rabbitts, T. H. (1991). Molecular cloning of putative odorant-binding and odorant-metabolizing proteins, *Biochemistry*, *30*, 10376-10382.

[60] Brigelius-Flohe, R. & Maiorino, M. (2013). Glutathione peroxidases, *Biochim Biophys Acta*, *1830*, 3289-3303.

[61] Utomo, A., Jiang, X., Furuta, S., Yun, J., Levin, D. S., Wang, Y. C., Desai, K. V., Green, J. E., Chen, P. L. & Lee, W. H. (2004). Identification of a novel putative non-selenocysteine containing phospholipid hydroperoxide glutathione peroxidase (NPGPx) essential for alleviating oxidative stress generated from polyunsaturated fatty acids in breast cancer cells, *J Biol Chem*, *279*, 43522-43529.

[62] Nguyen, V. D., Saaranen, M. J., Karala, A. R., Lappi, A. K., Wang, L., Raykhel, I. B., Alanen, H. I., Salo, K. E., Wang, C. C. & Ruddock, L. W. (2011). Two endoplasmic reticulum PDI peroxidases increase the efficiency of the use of peroxide during disulfide bond formation, *J Mol Biol*, *406*, 503-515.

[63] Wei, P. C., Hsieh, Y. H., Su, M. I., Jiang, X., Hsu, P. H., Lo, W. T., Weng, J. Y., Jeng, Y. M., Wang, J. M., Chen, P. L., Chang, Y. C., Lee, K. F., Tsai, M. D., Shew, J. Y. & Lee, W. H. (2012). Loss of the oxidative stress sensor NPGPx compromises GRP78 chaperone activity and induces systemic disease, *Mol Cell*, *48*, 747-759.

[64] Toppo, S., Vanin, S., Bosello, V. & Tosatto, S. C. (2008). Evolutionary and structural insights into the multifaceted glutathione peroxidase (Gpx) superfamily, *Antioxid Redox Signal*, *10*, 1501-1514.

[65] Yamada, Y., Limmon, G. V., Zheng, D., Li, N., Li, L., Yin, L., Chow, V. T., Chen, J. & Engelward, B. P. (2012). Major shifts in the spatio-temporal distribution of lung antioxidant enzymes during influenza pneumonia, *Plos one*, *7*, e31494.

[66] Vacanti, N. M. & Metallo, C. M. (2013). Exploring metabolic pathways that contribute to the stem cell phenotype, *Biochim Biophys Acta*, *1830*, 2361-2369.

[67] Ramm Sander, P., Hau, P., Koch, S., Schutze, K., Bogdahn, U., Kalbitzer, H. R. & Aigner, L. (2013). Stem cell metabolic and spectroscopic profiling, *Trends Biotechnol*, *31*, 204-213.

[68] Ito, K., Hirao, A., Arai, F., Matsuoka, S., Takubo, K., Hamaguchi, I., Nomiyama, K., Hosokawa, K., Sakurada, K., Nakagata, N., Ikeda, Y., Mak, T. W. & Suda, T. (2004). Regulation of oxidative stress by ATM is required for self-renewal of haematopoietic stem cells, *Nature*, *431*, 997-1002.

[69] Tothova, Z., Kollipara, R., Huntly, B. J., Lee, B. H., Castrillon, D. H., Cullen, D. E., McDowell, E. P., Lazo-Kallanian, S., Williams, I. R., Sears, C., Armstrong, S. A., Passegue, E., DePinho, R. A. & Gilliland, D. G. (2007). FoxOs are critical mediators of hematopoietic stem cell resistance to physiologic oxidative stress, *Cell*, *128*, 325-339.

[70] Chuikov, S., Levi, B. P., Smith, M. L. & Morrison, S. J. (2010). Prdm16 promotes stem cell maintenance in multiple tissues, partly by regulating oxidative stress, *Nat Cell Biol*, *12*, 999-1006.

[71] Aguilo, F., Avagyan, S., Labar, A., Sevilla, A., Lee, D. F., Kumar, P., Lemischka, I. R., Zhou, B. Y. & Snoeck, H. W. (2011). Prdm16 is a physiologic regulator of hematopoietic stem cells, *Blood*, *117*, 5057-5066.

[72] Horn, K. H., Warner, D. R., Pisano, M. & Greene, R. M. (2011). PRDM16 expression in the developing mouse embryo, *Acta Histochem*, *113*, 150-155.

[73] Kajimura, S., Seale, P., Kubota, K., Lunsford, E., Frangioni, J. V., Gygi, S. P. & Spiegelman, B. M. (2009). Initiation of myoblast to brown fat switch by a PRDM16-C/EBP-beta transcriptional complex, *Nature*, *460*, 1154-1158.

[74] De Haan, J. B., Crack, P. J., Flentjar, N., Iannello, R. C., Hertzog, P. J. & Kola, I. (2003). An imbalance in antioxidant defense affects cellular function: the pathophysiological consequences of a reduction in antioxidant defense in the glutathione peroxidase-1 (Gpx1) knockout mouse, *Redox Rep*, *8*, 69-79.

[75] Florian, S., Krehl, S., Loewinger, M., Kipp, A., Banning, A., Esworthy, S., Chu, F. F. & Brigelius-Flohe, R. (2010). Loss of GPx2 increases apoptosis, mitosis, and GPx1 expression in the intestine of mice, *Free Radic Biol Med*, *49*, 1694-1702.

[76] El Haddad, M., Jean, E., Turki, A., Hugon, G., Vernus, B., Bonnieu, A., Passerieux, E., Hamade, A., Mercier, J., Laoudj-Chenivesse, D. & Carnac, G. (2012). Glutathione peroxidase 3, a new retinoid target gene, is crucial for human skeletal muscle precursor cell survival, *J Cell Sci*, *125*, 6147-6156.

[77] Herault, O., Hope, K. J., Deneault, E., Mayotte, N., Chagraoui, J., Wilhelm, B. T., Cellot, S., Sauvageau, M., Andrade-Navarro, M. A., Hebert, J. & Sauvageau, G. (2012). A role for GPx3 in activity of normal and leukemia stem cells, *J Exp Med*, *209*, 895-901.

[78] Baliga, M. S., Wang, H., Zhuo, P., Schwartz, J. L. & Diamond, A. M. (2007). Selenium and GPx-1 overexpression protect mammalian cells against UV-induced DNA damage, *Biol Trace Elem Res*, *115*, 227-242.

[79] Liu, J., Du, J., Zhang, Y., Sun, W., Smith, B. J., Oberley, L. W. & Cullen, J. J. (2006). Suppression of the malignant phenotype in pancreatic cancer by overexpression of phospholipid hydroperoxide glutathione peroxidase, *Hum Gene Ther*, *17*, 105-116.

[80] Al-Taie, O. H., Uceyler, N., Eubner, U., Jakob, F., Mork, H., Scheurlen, M., Brigelius-Flohe, R., Schottker, K., Abel, J., Thalheimer, A., Katzenberger, T., Illert, B., Melcher, R. & Kohrle, J. (2004). Expression profiling and genetic alterations of the selenoproteins GI-GPx and SePP in colorectal carcinogenesis, *Nutr Cancer*, *48*, 6-14.

[81] Chiu, S. T., Hsieh, F. J., Chen, S. W., Chen, C. L., Shu, H. F. & Li, H. (2005). Clinicopathologic correlation of up-regulated genes identified using cDNA microarray and real-time reverse transcription-PCR in human colorectal cancer, *Cancer Epidemiol Biomarkers Prev*, *14*, 437-443.

[82] Mork, H., Scheurlen, M., Al-Taie, O., Zierer, A., Kraus, M., Schottker, K., Jakob, F. & Kohrle, J. (2003). Glutathione peroxidase isoforms as part of the local antioxidative defense system in normal and Barrett's esophagus, *Int J Cancer*, *105*, 300-304.

[83] Serewko, M. M., Popa, C., Dahler, A. L., Smith, L., Strutton, G. M., Coman, W., Dicker, A. J. & Saunders, N. A. (2002). Alterations in gene expression and activity during squamous cell carcinoma development, *Cancer Res*, *62*, 3759-3765.

[84] Woenckhaus, M., Klein-Hitpass, L., Grepmeier, U., Merk, J., Pfeifer, M., Wild, P., Bettstetter, M., Wuensch, P., Blaszyk, H., Hartmann, A., Hofstaedter, F. & Dietmaier, W. (2006). Smoking and cancer-related gene expression in bronchial epithelium and non-small-cell lung cancers, *J Pathol*, *210*, 192-204.

[85] Lee, O. J., Schneider-Stock, R., McChesney, P. A., Kuester, D., Roessner, A., Vieth, M., Moskaluk, C. A. & El-Rifai, W. (2005). Hypermethylation and loss of expression of glutathione peroxidase-3 in Barrett's tumorigenesis, *Neoplasia*, *7*, 854-861.

[86] Peng, D.F., Razvi, M., Chen, H., Washington, K., Roessner, A., Schneider-Stock, R. & El-Rifai, W. (2009). DNA hypermethylation regulates the expression of members of the Mu-class glutathione S-transferases and glutathione peroxidases in Barrett's adenocarcinoma, *Gut*, *58*, 5-15.

[87] He, Y., Wang, Y., Li, P., Zhu, S., Wang, J. & Zhang, S. (2011). Identification of GPX3 epigenetically silenced by CpG methylation in human esophageal squamous cell carcinoma, *Dig Dis Sci*, *56*, 681-688.

[88] Falck, E., Karlsson, S., Carlsson, J., Helenius, G., Karlsson, M. & Klinga-Levan, K. (2010). Loss of glutathione peroxidase 3 expression is correlated with epigenetic mechanisms in endometrial adenocarcinoma, *Cancer Cell Int*, *10*, 46.

[89] Yu, Y. P., Yu, G., Tseng, G., Cieply, K., Nelson, J., Defrances, M., Zarnegar, R., Michalopoulos, G. & Luo, J. H. (2007). Glutathione peroxidase 3, deleted or methylated in prostate cancer, suppresses prostate cancer growth and metastasis, *Cancer Res*, *67*, 8043-8050.

[90] Cejas, P., Garcia-Cabezas, M. A., Casado, E., Belda-Iniesta, C., De Castro, J., Fresno, J. A., Sereno, M., Barriuso, J., Espinosa, E., Zamora, P., Feliu, J., Redondo, A., Hardisson, D. A., Renart, J. & Gonzalez-Baron, M. (2007). Phospholipid hydroperoxide glutathione peroxidase (PHGPx) expression is downregulated in poorly differentiated breast invasive ductal carcinoma, *Free Radic Res*, *41*, 681-687.

[91] Heirman, I., Ginneberge, D., Brigelius-Flohe, R., Hendrickx, N., Agostinis, P., Brouckaert, P., Rottiers, P. & Grooten, J. (2006). Blocking tumor cell eicosanoid synthesis by GP x 4 impedes tumor growth and malignancy, *Free Radic Biol Med*, *40*, 285-294.

[92] Gyorffy, B., Lanczky, A., Eklund, A. C., Denkert, C., Budczies, J., Li, Q. & Szallasi, Z. (2010). An online survival analysis tool to rapidly assess the effect of 22,277 genes on breast cancer prognosis using microarray data of 1,809 patients, *Breast Cancer Res Treat*, *123*, 725-731.

[93] Liu, M., Wang, G., Gomez-Fernandez, C. R. & Guo, S. (2012). GREB1 Functions as a Growth Promoter and Is Modulated by IL6/STAT3 in Breast Cancer, *Plos one*, *7*, e46410.

[94] Nguyen, L. V., Vanner, R., Dirks, P. & Eaves, C. J. (2012). Cancer stem cells: an evolving concept, *Nat Rev Cancer*, *12*, 133-143.

[95] Guo, S., Liu, M., Wang, G., Torroella-Kouri, M. & Gonzalez-Perez, R. R. (2012). Oncogenic role and therapeutic target of leptin signaling in breast cancer and cancer stem cells, *Biochim Biophys Acta*, *1825*, 207-222.

[96] Parashurama, N., Lobo, N. A., Ito, K., Mosley, A. R., Habte, F. G., Zabala, M., Smith, B. R., Lam, J., Weissman, I. L., Clarke, M. F. & Gambhir, S. S. (2012). Remodeling of endogenous mammary epithelium by breast cancer stem cells, *Stem Cells*, *30*, 2114-2127.

[97] Medema, J. P. (2013). Cancer stem cells: The challenges ahead, *Nat Cell Biol*, *15*, 338-344.

[98] Pang, L. Y. & Argyle, D. J. (2009). Using naturally occurring tumours in dogs and cats to study telomerase and cancer stem cell biology, *Biochim Biophys Acta*, *1792*, 380-391.

[99] Kakarala, M. & Wicha, M. S. (2008). Implications of the cancer stem-cell hypothesis for breast cancer prevention and therapy, *J Clin Oncol*, *26*, 2813-2820.

[100] Clarke, M. F. (2005). Self-renewal and solid-tumor stem cells, *Biol Blood Marrow Transplant*, *11*, 14-16.

[101] Clevers, H. (2011). The cancer stem cell: premises, promises and challenges, *Nat Med*, *17*, 313-319.

[102] Diehn, M., Cho, R. W., Lobo, N. A., Kalisky, T., Dorie, M. J., Kulp, A. N., Qian, D., Lam, J. S., Ailles, L. E., Wong, M., Joshua, B., Kaplan, M. J., Wapnir, I., Dirbas, F. M., Somlo, G., Garberoglio, C., Paz, B., Shen, J., Lau, S. K., Quake, S. R., Brown, J. M., Weissman, I. L. & Clarke, M. F. (2009). Association of reactive oxygen species levels and radioresistance in cancer stem cells, *Nature*, *458*, 780-783.

[103] Ye, X. Q., Li, Q., Wang, G. H., Sun, F. F., Huang, G. J., Bian, X. W., Yu, S. C., Qian, G. S. (2011). Mitochondrial and energy metabolism-related properties as novel indicators of lung cancer stem cells, *Int J Cancer*, *129*, 820-831.

[104] Schulze-Bergkamen, H. & Krammer, P. H. (2004). Apoptosis in cancer--implications for therapy, *Semin Oncol*, *31*, 90-119.

[105] Dell'Eva, R., Pfeffer, U., Vene, R., Anfosso, L., Forlani, A., Albini, A. & Efferth, T. (2004). Inhibition of angiogenesis in vivo and growth of Kaposi's sarcoma xenograft tumors by the anti-malarial artesunate, *Biochem Pharmacol*, *68*, 2359-2366.

[106] Haraguchi, N., Ishii, H., Mimori, K., Tanaka, F., Ohkuma, M., Kim, H. M., Akita, H., Takiuchi, D., Hatano, H., Nagano, H., Barnard, G. F., Doki, Y. & Mori, M. (2010). CD13 is a therapeutic target in human liver cancer stem cells, *J Clin Invest*, *120*, 3326-3339.

[107] Christ, B., Stock, P. & Dollinger, M. M. (2011). CD13: Waving the flag for a novel cancer stem cell target, *Hepatology*, *53*, 1388-1390.

[108] Kim, H. M., Haraguchi, N., Ishii, H., Ohkuma, M., Okano, M., Mimori, K., Eguchi, H., Yamamoto, H., Nagano, H., Sekimoto, M., Doki, Y. & Mori, M. (2012). Increased CD13 expression reduces reactive oxygen species, promoting survival of liver cancer stem cells via an epithelial-mesenchymal transition-like phenomenon, *Ann Surg Oncol*, *19* Suppl 3, S539-548.

[109] Dong, C., Yuan, T., Wu, Y., Wang, Y., Fan, T. W., Miriyala, S., Lin, Y., Yao, J., Shi, J., Kang, T., Lorkiewicz, P., St Clair, D., Hung, M. C., Evers, B. M. & Zhou, B. P. (2013). Loss of FBP1 by Snail-mediated repression provides metabolic advantages in basal-like breast cancer, *Cancer Cell*, *23*, 316-331.

[110] Park, I. J., Lee, Y. K., Hwang, J. T., Kwon, D. Y., Ha, J. & Park, O. J. (2009). Green tea catechin controls apoptosis in colon cancer cells by attenuation of H2O2-stimulated COX-2 expression via the AMPK signaling pathway at low-dose H2O2, *Ann N Y Acad Sci*, *1171*, 538-544.

[111] Du, G. J., Lin, H. H., Xu, Q. T. & Wang, M. W. (2006). Bcl-2 switches the type of demise from apoptosis to necrosis via cyclooxygenase-2 upregulation in HeLa cell induced by hydrogen peroxide, *Cancer Lett*, *232*, 179-188.

[112] Cesario, A., Rocca, B. & Rutella, S. (2011). The interplay between indoleamine 2,3-dioxygenase 1 (IDO1) and cyclooxygenase (COX)-2 in chronic inflammation and cancer, *Curr Med Chem*, *18*, 2263-2271.

[113] Harris, R. E. (2007). Cyclooxygenase-2 (cox-2) and the inflammogenesis of cancer, *Subcell Biochem*, *42*, 93-126.

[114] Sharma, V., Dixit, D., Ghosh, S. & Sen, E. (2011). COX-2 regulates the proliferation of glioma stem like cells, *Neurochem Int*, *59*, 567-571.

[115] Wang, K. H., Kao, A. P., Chang, C. C., Lee, J. N., Hou, M. F., Long, C. Y., Chen, H. S. & Tsai, E. M. (2010). Increasing CD44+/CD24(-) tumor stem cells, and upregulation of COX-2 and HDAC6, as major functions of HER2 in breast tumorigenesis, *Mol Cancer*, *9*, 288.

[116] Singh, B., Cook, K. R., Vincent, L., Hall, C. S., Berry, J. A., Multani, A. S. & Lucci, A. (2008). Cyclooxygenase-2 induces genomic instability, BCL2 expression, doxorubicin resistance, and altered cancer-initiating cell phenotype in MCF7 breast cancer cells, *J Surg Res*, *147*, 240-246.

[117] Wang, D. & Dubois, R. N. (2006). Prostaglandins and cancer, *Gut*, *55*, 115-122.

[118] Greenhough, A., Smartt, H. J., Moore, A. E., Roberts, H. R., Williams, A. C., Paraskeva, C. & Kaidi, A. (2009). The COX-2/PGE2 pathway: key roles in the hallmarks of cancer and adaptation to the tumour microenvironment, *Carcinogenesis*, *30*, 377-386.

[119] Jain, S., Chakraborty, G., Raja, R., Kale, S. & Kundu, G. C. (2008). Prostaglandin E2 regulates tumor angiogenesis in prostate cancer, *Cancer Res*, *68*, 7750-7759.
[120] Shao, J., Sheng, G. G., Mifflin, R. C., Powell, D. W. & Sheng, H. (2006). Roles of myofibroblasts in prostaglandin E2-stimulated intestinal epithelial proliferation and angiogenesis, *Cancer Res*, *66*, 846-855.
[121] Castellone, M. D., Teramoto, H., Williams, B. O., Druey, K. M. & Gutkind, J. S. (2005). Prostaglandin E2 promotes colon cancer cell growth through a Gs-axin-beta-catenin signaling axis, *Science*, *310*, 1504-1510.
[122] Ishimoto, T., Oshima, H., Oshima, M., Kai, K., Torii, R., Masuko, T., Baba, H., Saya, H. & Nagano, O. (2010). CD44+ slow-cycling tumor cell expansion is triggered by cooperative actions of Wnt and prostaglandin E2 in gastric tumorigenesis, *Cancer Sci*, *101*, 673-678.
[123] C.H.-Y., Dayem, A. A., Kim, J-H. & Cho, S-G. (2010). Role of Oxidative Stress in Stem, Cancer, and Cancer Stem Cells *Cancers*, *2*(2), 859-884
[124] Shi, X., Zhang, Y., Zheng, J. & Pan, J. (2012). Reactive oxygen species in cancer stem cells, *Antioxid Redox Signal*, *16*, 1215-1228.
[125] Dokic, I., Hartmann, C., Herold-Mende, C. & Regnier-Vigouroux, A. (2012). Glutathione peroxidase 1 activity dictates the sensitivity of glioblastoma cells to oxidative stress, *Glia*, *60*, 1785-1800.
[126] Li, H. J., Reinhardt, F., Herschman, H. R. & Weinberg, R. A. (2012). Cancer-stimulated mesenchymal stem cells create a carcinoma stem cell niche via prostaglandin E2 signaling, *Cancer Discov*, *2*, 840-855.
[127] Lubos, E., Kelly, N. J., Oldebeken, S. R., Leopold, J. A., Zhang, Y. Y., Loscalzo, J. & Handy, D. E. (2011). Glutathione peroxidase-1 deficiency augments proinflammatory cytokine-induced redox signaling and human endothelial cell activation, *J Biol Chem*, *286*, 35407-35417.
[128] Felice, F., Lucchesi, D., di Stefano, R., Barsotti, M. C., Storti, E., Penno, G., Balbarini, A., Del Prato, S. & Pucci, L. (2010). Oxidative stress in response to high glucose levels in endothelial cells and in endothelial progenitor cells: evidence for differential glutathione peroxidase-1 expression, *Microvasc Res*, *80*, 332-338.
[129] Galasso, G., Schiekofer, S., Sato, K., Shibata, R., Handy, D. E., Ouchi, N., Leopold, J. A., Loscalzo, J. & Walsh, K. (2006). Impaired angiogenesis in glutathione peroxidase-1-deficient mice is associated with endothelial progenitor cell dysfunction, *Circ Res*, *98*, 254-261.
[130] Zhu, T. S., Costello, M. A., Talsma, C. E., Flack, C. G., Crowley, J. G., Hamm, L. L., He, X., Hervey-Jumper, S. L., Heth, J. A., Muraszko, K. M., DiMeco, F., Vescovi, A. L. & Fan, X. (2011). Endothelial cells create a stem cell niche in glioblastoma by providing NOTCH ligands that nurture self-renewal of cancer stem-like cells, *Cancer Res*, *71*, 6061-6072.
[131] Gursel, D. B., Berry, N. & Boockvar, J. A. (2012). The contribution of Notch signaling to glioblastoma via activation of cancer stem cell self-renewal: the role of the endothelial network, *Neurosurgery*, *70*, N19-21.

In: Peroxidases
Editors: Lucas Bogaert and Noah Coppens
ISBN: 978-1-62808-261-6

Chapter V

Biochemical Responses of Aquatic Macrophytes to Two Abiotic Stresses

Takashi Asaeda**, Tanjeena Zaman and Md H. Rashid
Department of Environmental Science, Saitama University,
Sakura, Saitama, Japan

Abstract

Exposure to stress can lead to the disruption of cellular, molecular and physiological processes in plants. Plants generate ROS to cope with different forms of abiotic stresses. However, if a plant is exposed to stress conditions for a considerable period, the excess generation of ROS affects its radical scavenging ability in the long run. Thus the effects of environmental variability of a plant are reflected in its growth, reproduction and colonization. Aquatic plants are considered unchangeable biological filters and play an important role in the proper functioning of aquatic ecosystem. They often encounter abiotic stress of varying magnitudes and acquire with different adaptive characters. In this chapter, we will be focusing on the oxidative responses of aquatic macrophytes to various abiotic stresses with a special concern on flow velocity and redox states.

General Introduction

Aquatic plant habitats are characterized by high spatial and temporal variations of physical and chemical parameters which are considered as heterogeneous environments. Under natural conditions, plants frequently encounter combinations of stress factors like mechanical and resource stress. In such conditions, plant growth is altered in a complex interactive manner that cannot be simply predicted from the responses to each stress factor when considered independently. Thus, aquatic plants specially submerged macrophytes

* Corresponding author: Dr. Takashi Asaeda, Department of Environmental Science, Saitama University, 255 Shimo-okubo, Sakura, Saitama, 338-8570, Japan, E. mail: asaeda@mail.saitama-u.ac.jp.

frequently encounter with various stress factors. Consequently, the individual ability to tolerate multiple stresses through morphological adjustments is a major feature that determines species survival and colonization, and hence the ecological breadth of the species (Chapin et al., 1987). The abiotic factors include organic substances, pH, temperature, alkalinity and hardness, inorganic ligands, interactions, sediments, redox status etc. The excess generation of ROS in plant cell is the biochemical response regarding any kind of stress (Gill and Tuteja, 2010). Free radicals and ROS are generated naturally in the chloroplasts, mitochondria, endoplasmic reticulum, peroxisomes, glyoxysomes, plasma membrane, and apoplasm of plant cells (Wise and Naylor, 1987). Free radical reactive intermediates can react directly or indirectly with molecular oxygen to form reactive oxygen species (ROS), such as superoxide anion (O_2^-), hydrogen peroxide (H_2O_2), and hydroxyl radicals (OH^-) (Navrot et al., 2007). Chloroplasts are especially prone to generate ROS during photosynthesis as the internal O_2 concentrations are high (Asada and Takahashi, 1987). Electrons leaked from electron transport chains can react with O_2 during normal aerobic metabolism to produce ROS. Hydroxyl radicals (OH^-), produced from O_2^- and H_2O_2 by cell wall peroxidase, have been observed to be involved in cleaving cell wall polymers and to act as a cell wall loosening agent, thus mediating auxin induced extension growth under normal growth conditions (Liszkay et al. 2003). Once produced, O_2^- will rapidly dismutate, either enzymatically or nonenzymatically, to yield H_2O_2 and O_2. In addition, H_2O_2 and O_2^- may interact in the presence of Fe^{2+} to yield the highly reactive OH^- (Imlay and Linn, 1988). Free radicals can be generated in plants when a nonradical gains or loses an electron, or receives excitation energy from a photoactivated pigment. Most radicals are unstable and are therefore short lived; half-lives in the order of a microsecond or a nanosecond are common. Stability is restored when two radicals meet and share their unpaired electrons in a covalent bond. However, when a free radical reacts with an organic molecule such as an unsaturated fatty acid in a membrane bilayer, a new radical results, and chain reactions are established. Such reactions can eventually lead to oxidative injury.

H_2O_2 has also been shown to act as a key regulator in a broad range of physiological processes, such as senescence, photorespiration and photosynthesis, stomatal movement, cell cycle and growth and development (Moustakas et al., 1994). H_2O_2 is starting to be accepted as a second messenger for signals generated by means of ROS because of its relatively long life and high permeability across membranes (Schutzendubel et al., 2001). When ROS generation exceeds the capacity of the cellular antioxidants, it will cause oxidative stress and significant oxidative damage to a plant. These cytotoxic ROS can strongly disrupt normal metabolism through oxidative damage of chlorophyll, lipids, protein, and nucleic acids (Herbinger et al., 2002). To maintain its functional and structural integrity, a plant organism has to be resistant towards unfavourable factors. Each organism has a unique range of genetically determined and phylogenetically adapted physiological resistance within which a factor affecting it is tolerable. If a stress factor surpasses this range the plant has to trigger additional energy and physiological–biochemical mechanisms to survive under unfavourable conditions. To protect them against oxidative stress, plant cells produce both antioxidant enzymes, such as superoxide dismutase (SOD), catalase (CAT), peroxidase (POD), glutathione-S-transferase (GST), and nonenzymatic antioxidants such as glutathione and ascorbate (Foyer and Noctor, 2005). Table-1 shows the different (Mehlhorn et al., 1996) enzymes responsible for ROS scavenging and detoxification.

Oxidative stress may indirectly affect mitotic activation by enhancing effects on the cell growth cycle. Under stress exposure, H_2O_2 and IAA have antagonistic effects on cell cycle progression (Kovtun et al. 2000). High levels of reactive oxygen species under stress conditions may block the cell cycle at specific check points (Pasternak et al. 2005; Van Breusegem et al. 2001). The adaptive response of various crop plants and their antioxidative response to various stress factors have been studied for the last three decades (see for example, Mittler et al., 2002; Gill and Tuteja 2010, etc.).

The effect of different abiotic factors on aquatic macrophyte specially reduced environment and turbulence flow are very scanty (see for example, DeLaune et al., 1999; Jackson et al., 1993; Ellawala et al., 2011 a,b). In this chapter, we will be focusing on the oxidative responses of aquatic macrophytes to various abiotic stresses with a special concern on flow velocity and redox states.

Table 1. ROS scavenging and detoxifying enzymes after Mehlhorn et al. (1996)

Enzyme	EC umber	Reaction catalyzed
Superoxide dismutase	1.15.1.1	$O_2^{\cdot-} + O_2^{\cdot-} + 2H^+ \leftrightarrow 2H_2O_2 + O_2$
Catalase	1.11.1.6	$2H_2O_2 \leftrightarrow O_2 + 2H_2O$
Glutathione peroxidase	1.11.1.12	2GSH + PUFA-OOH $\leftrightarrow$ GSSG + PUFA + $2H_2O$
Glutathione S-transferases	2.5.1.18	RX + GSH $\leftrightarrow$ HX + R-S-GSH*
Phospholipid-hydroperoxide glutathione peroxidase	1.11.1.9	2GSH + PUFA-OOH (H_2O_2) $\leftrightarrow$ GSSG + $2H_2O$**
Ascorbate peroxidase	1.11.1.11	AA + H_2O_2 $\leftrightarrow$ DHA + $2H_2O$
Guaiacol type peroxidase	1.11.1.7	Donor + H_2O_2 $\leftrightarrow$ oxidized donor + $2H_2O$***
Monodehydroascorbate reductase	1.6.5.4	NADH + 2MDHA $\leftrightarrow$ NAD+ + 2AA
Dehydroascorbate reductase	1.8.5.1	2GSH + DHA $\leftrightarrow$ GSSG + AA
Glutathione reductase	1.6.4.2	NADPH + GSSG $\leftrightarrow$ NADP+ + 2GSH

*R may be an aliphatic, aromatic or heterocyclic group; X may be a sulfate, nitrite or halide group.
**Reaction with H_2O_2 is slow.
*** AA acts as an electron donor.

Antioxidative Response Due to Flow Turbulence

Mechanical Stress Due to Water Flow

In aquatic habitats (both lentic and lotic), mechanical stressors result from an exposure to water flow. Water movement influences the distribution and metabolism of submerged macrophytes. (Sabbatini et al., 1998). Submerged macrophytes in flowing waters experience drag and lift forces that are caused by the flow velocity (Madsen et al., 2001; Sand-Jensen, 2008). In shallow lakes, wind induced currents and waves create forces on the macrophytes, and the magnitude of these forces depends on depth and fetch (Schutten et al., 2005). In wetlands and shallow lakes, where there is no mean flow, wind-induced turbulence is the major factor that affects the bottom boundary layer (Pujol et al. 2010). Aquatic vegetation affects the mean velocity of flow, the turbulence intensity and diffusion. In addition, aquatic plants convert mean kinetic energy into turbulent kinetic energy at the scale of stems and branches (Nepf, 1999).

In general, small scale water movement has positive effects on nutrient uptake and metabolism due to the reduction of thickness of diffusion boundary layer around leaves (Madsen et al., 1993; Nishihara and Ackerman, 2006). However, water movement does not necessarily enhance nutrient uptake; it can inhibit plant physiological processes becoming a stress to the plants. Reduction of growth, variations of growth hormone concentration and development of oxidative stress have been observed when plants are exposed to flow and/or turbulence (Ke and Li, 2006; Ellawala et al., 2011a,b). In addition, reduction of the rate of photosynthesis, metabolism and reduction of nutrient uptake have also been observed in many aquatic plants, when they are exposed to water movement of any kind (Madsen et al., 1993; Ellawala et al., 2011a).

The effect of turbulence on a plant's metabolism and physiology is species specific (Asaeda et al. 2005, 2010). Natural water flows can be divided into two components: mean flow and turbulence. The main flow is relatively steady and uni-directional, where shoots of submerged plants become aligned with the direction of flow. Therefore, shoots receive shear stress from the water, which stretches the plant body longitudinally. Mean flow favors the growth of submerged plants by thinning their boundary layer thickness on the surface and increasing the CO_2 and nutrient flux to the photosynthetic organ (Fonseca and Kenworth, 1987), while ethylene escapes from the submerged tissues (Musgrave et al., 1972), increasing the IAA level (Ke and Li, 2006). By contrast, turbulence alters its magnitude and direction constantly. The frequency of the fluctuation is composed of a relatively wide range of components from high to low frequencies. A plant does not follow a high frequency type of motion and the velocity gap between the plant body and fluid produces eddies behind it, which causes drag on the plant body. By lowering the frequency, plants can follow the direction of the flow more easily, where the stress on the plant body changes causing a shearing force on the surface. However, as the direction of flow changes within a short period of time, it does not become stronger. Submerged plants illicit a vacillating type of motion under water, therefore, the movement of the plant body is measured to be less than that of fluid.

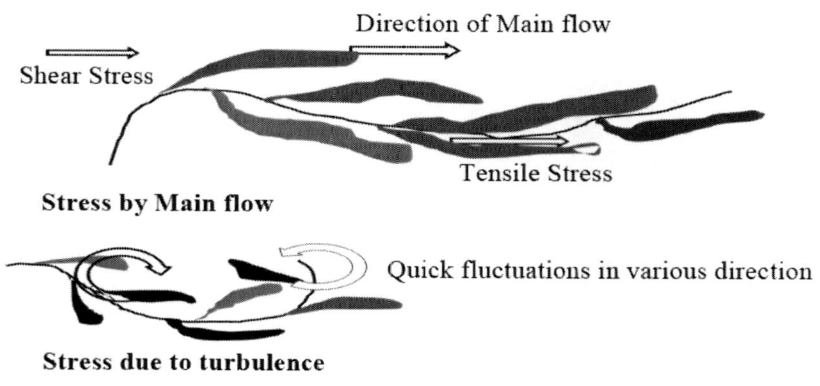

Figure 1. Schematic diagram showing stress due to main flow and turbulence, encountered on submerged macrophytes.

Floating leaved plants, on the other hand, are subjected to a wave type of motion on the surface of the water. As their leaves float with a higher amount of buoyancy, they follow the up and down movement of the water surface. Thus, their stems periodically receive tensile stress. It seems to have high impacts on the plant cells. The stresses of these flow components on submerged plants are different from each other. For example, mean flow acts as a tensile shear stress on the body of the plant, while turbulence flagellates it in an irregular way (Figure 1). Therefore, the responses of the plants to the stresses are assumed to be different as well.

Morphological Adaptation

Plants use morphogenic responses to decrease stress exposure (Bal et al., 2011). SIMRs (Stress induced morphogenic responses) comprise three components: (a) inhibition of cell elongation, (b) localized stimulation of cell division and (c) alterations in cell differentiation status (Potters, 2009). Plants use morphogenic responses to decrease stress exposure. The similarities between morphogenicre sponses induced by distinct stresses reflect common molecular effects, such as increased ROS-production and altered auxin metabolism.

Stress can boost the stress tolerance of the plant through the induction of acclimation responses. For example most submerged plants have flexible shoots and leaves with streamlined morphology proportional to the drag force (Sand-Jensen, 2003). Morphological patterns that support to prevent plant damage or uprooting in turbulent environments differ among different species. For example, species like *Vallisneria* have long, ribbon-like leaves to minimize tearing (Doyle 2001). Free floating leaved plants had a larger leaf diameter than completely submerged leaves plants within low velocity streams (Boeger and Poulson, 2002). Flexible plants minimize the forces experienced in moving water by bending over and forming a shielding canopy that reduces the frontal area (Sand-Jensen 2003). Ellawala et al., (2011 b) observed decreased shoot elongation of *E. densa* is a morphogenetic response that results in less drag on the shoots. Though some of these modifications reduce damage to mature plants, in the early stages of growth are more vulnerable to damage (Doyle 2001), as the benefit of canopy formation is not yet available. Downing-Kunz and Stacey (2012) showed the presence of the root canopy in free floating plant (*E. crassipes)* caused deflection of flow around and reduced velocities within the canopy. Water movement induced mechanical stress may have a major impact on the survival of winter buds, turions or seedlings, which grow as individuals (Doyle 2001; Ke and Li 2006).

Cellular and Antioxidative Response

An exposure to this kind of stress (flow turbulence) can lead to the disruption of cellular and molecular processes. It is reported that a turbulent jet affects plant cells and may induce oxidative stress, as do many other abiotic stresses (Pasternak et al., 2005). Mechanical stress generated by water movement has been observed to reduce growth hormone concentrations and to retard growth due to accumulation of reactive oxygen molecules (Ellawala et al., 2011b). Photosynthetic carbon uptake has been observed to reduce when the plants are exposed to current velocity induced movements (Madsen et al., 1993). At low velocity, the reduction in thickness of the boundary layer surrounding the macrophyte leaves and an increase in aeration of the water with atmospheric carbon dioxide (Fox, 1996) improves photosynthetic rates. However at higher velocities, photosynthesis can be inhibited by a current (Madsen et al., 1993) and also by reduction of light penetration (Westlake, 1967). Plant hormones play a major role in the regulation of developmental processes and in signaling networks involved in responses to biotic and abiotic stress (Bari and Jones, 2009). Hormonal variations play a major role in the response of plants to environmental conditions. There is an emerging idea that auxin and oxidative stress are complementary during the stress response and stress induced reorientation of growth (Pasternak et al., 2005). The response of aquatic plants to turbulence also induce changes in plant hormones as explored recently by various researchers (for example see; Ke and Li 2006; Ellawala et al., 2011 a, b). Indole acetic acid (IAA) is one of the major growth hormones related to shoot elongation and cell division. Reduced plant growth under stress conditions allows the plant to conserve energy and initiate the appropriate defense mechanisms necessary for survival (Van Breusegem et al., 2001). The regulation of auxin transport and auxin-dependent activities by H_2O_2 may shift energy utilization toward stress protection and survival (Kovtun et al., 2000).

Ellawala et al., (2011 a, b) found a decrease in the SER (Shoot elongation rate) in all experimental submerged plants under high turbulence, which might be the alternation of phytohormones like IAA and cytokinin (CK) as found in terrestrial plants (Chehab et al., 2009). These results were in accordance with the earlier studies conducted with flow turbulence in the higher aquatic plants (Puijalon et al., 2005). Plants exposed to low turbulence velocity showed a slight increase in shoot length, which supports the hypothesis that mechanical vibration reduces the activity of IAA oxidase and enhances super oxidase, then the growth rate was enhanced with some vibration (Yang et al., 2002). However, long-term exposure to high turbulence may have caused the plant to reduce IAA production, hence reducing IAAO activity and resulting in adaptation to the environment. Ellawala et al., (2011 a) found the increment of IAA in *P. crispus* was observed that may have promoted by auxin-induced cell elongation and hence increased growth under the situations favorable to nutrient uptake during exposure to turbulence, as observed by Nielsen et al. (2006). For other species the decrement of IAA in plants under different turbulence velocities as IAA was suggested to prevent the loss of chlorophylls through aging of chloroplast (Misra and Biswal, 1980). Catalase (CAT) is a high capacity but low affinity enzyme which destroys hydrogen peroxide. Ellawala et al., (2012) found the increased activities of CAT and APX which also confirms the development of oxidative stress in the plants exposed to high and medium turbulence compared with low turbulence and the control.

Over production of ROS increases the content of malondialdehyde (MDA) .The content of MDA has been considered an indicator of oxidative damage (Møller et al., 2007). MDA is considered as a suitable marker for membrane lipid peroxidation. A decrease in membrane stability reflects the extent of lipid peroxidation caused by ROS. Furthermore, lipid peroxidation is an indicator of the prevalence of free radical reaction in tissues. Moreover, oxygen uptake loading on the tissues as both processes generate reactive oxygen species, particularly H_2O_2 that produced at very high rates by the glycollate oxidase reaction in the peroxisomes in photorespiration. Previous studies regarding flow turbulence on aquatic plants showed that with the increasing turbulence velocity, the MDA concentration increased in plants. As *M. spicatum* is completely submerged species, the whole body was affected by the shear stress, whereas for floating leaved aquatic plants, leaf received more stress than the stem (Schutten et al., 2004). The leaves of all of the plants had high endogenous H_2O_2 concentrations when exposed to high turbulence (Ellawala et al., 2011). H_2O_2 is a necessary substrate for the cell wall stiffening process catalyzed by peroxided, but high concentration of H_2O_2 inhibits auxin mediated growth and H_2O_2 causes cross-linking of cell wall polymers, which ultimately followed to leaf senescence. Increased POD and IAAO activity as a result of exposure to high and medium turbulence indicates a triggered amount of oxidative stress. CKX activity of the plants, which can be used to measure plant CK (cytokinin) catabolism (Libreros-Minotta and Tipton, 1995), was found to increase with the elevation of turbulence velocity. Higher CKX activity was also found in plants exposed to high turbulence, which indicates that CK may be catabolized in a higher rate under these conditions.

Excessive levels of H_2O_2 over a long period may exceed the antioxidant capacity of the plant and damage the cellular photosynthetic machinery (Van Breusegem et al., 2001). Plants lost biomass through leaf senescence in high turbulence. During periods of severe stress, the H_2O_2 scavenging system may have been saturated, as the antioxidant capacity of the plants might have been exceeded (Van Breusegem et al. 2001). It can therefore be hypothesized that the observed H_2O_2 accumulation in tissues may promote the loss of leaves. Meanwhile, the emerging new leaves were shorter and thinner, and fewer leaves emerged. High IAAO and POD activities are known to catabolize IAA (Siegel 1993; Hiraga et al. 2001) and reduce auxin-induced growth; this may cause the above observation. Some studies have suggested that high H_2O_2 concentrations in the tissues alter the levels and distribution of IAA, the principal form of auxin in plants. The variation in the H_2O_2 concentration with turbulence velocity was lower in *E. nuttallii* than in *V. spiralis*, and it can therefore be hypothesized that the antioxidant capacity of *E. nuttallii* may be more capable of tolerating increased H_2O_2. Supporting this observation, *P. natans* and *P. pectinatus* have been observed to grow longer when exposed to currents in the field, while in the same study under the same conditions, other species were observed to grow less when exposed to currents and waves (Szmeja and Galka 2008). POD is believed to be involved in the lignification of secondary cell walls by catalyzing the polymerization of phenolics into lignin, and in the forming of covalent cross-links between ferulylated polysaccharides, hydroxyproline-rich glycoproteins and lignin (Zhang et al., 2009). The increased plant growth rate is associated with the decreased H_2O_2 and MDA content. Mohan and Hosetti (1997) suggested that the increase of peroxidase activity may be an effect of accelerated senescence, connected with enhanced formation of hydrogen peroxide (H_2O_2) or secondary metabolites such as phenolic compounds. The regulation of auxin transport and auxin-dependent activities by H_2O_2 may shift energy utilization toward stress protection and survival (Kovtun et al. 2000).

Antioxidative Response Due to Low Redox Status of Sediment

Plants in flooded soil or wetland come across with different stress factors among them the principal stress factor is biochemical reduction, the intensity of which is measured as redox potential (Lissner et al., 2003). Reduction of redox potential dramatically alters the chemistry and microbial metabolism at the sediment-water interface; this is a common phenomenon in water-logged soil. The reducing chemicals and biogeochemical processes induced by oxygen depletion accelerate the sediments oxygen demand, lowering its redox potential to values as -300mV (Delaune et al., 1990). According to Devai and DeLaune, (1995) redox potentials of soils can range from -300 to + 700 mV and anaerobic soils exhibit redox potentials from + 350 mV to as low as - 250 to -300 mV. Soils tend to undergo a series of sequential redox reactions in a homogenous environment when soil redox status changes from aerobic (high E_H) to anaerobic (low E_H) conditions and vice versa. Major reactions include, in order of E_H from high to low, nitrification, denitrification, manganic manganese [Mn (IV)] reduction, ferric iron [Fe (III)] reduction, sulfate (SO_4^{2-}) reduction, and methanogenesis (table-2).

With the reduction of oxides and microbial activity elements that are fixed in them such as phosphorus (P), molybdenum (Mo), cobalt (Co), copper (Cu), Zn are often transformed to a more mobile and plant available form under anoxic condition. Lower sediment pH under mildly oxic redox conditions increase the bioavailability of Al, Cu, Fe, Mn and Zn to rooted aquatic plants many of which considered as highly phytotoxic (Pezeshki and DeLaune, 2012). Low oxygen concentration in the environment is a natural and common environmental challenge that plants face throughout their life. A lack of O2 may negatively affect plant metabolism, including nitrogen uptake and assimilation (Gibbs and Greenway, 2003) at the same time a decrease in adenylate energy charge, cytoplasmic acidification, anaerobic fermentation, elevation in cytosolic Ca+2 concentration, changes in redox state, a membrane barrier function and inhibited nutrient uptake and maintenance are the main features caused by lack of oxygen (reviewed by Crawford 1992; Drew, 1997).

Table 2. Range of redox potentials required to reduce oxidized forms of the various redox couples in soil and wetland environments

Redox Couple in Wetland Soils (*Ox→ Red*)	Range of Measured Redox Potential (mV)
$O_2 \rightarrow H_2O$	+400 to +350
$NO_3- \rightarrow N_2$	+250 to +200
$Mn_{4+} \rightarrow Mn_2+$	+200 to +150
Fe3+→Fe2+	-25 to -75
$SO_42- \rightarrow S2-$	-125 to -175
$CO_2 \rightarrow CH_4$	-200 to -250

Source: Patrick, et al., 1996

Anoxic conditions results in a number of ecological processes that can degrade water quality. Under reducing conditions, NH_4^+ is the available form of N in the environment. Several previous studies have shown that NH_4^+ enrichment can directly cause the decline in aquatic plant populations in natural waters (Moore et al., 1996; Goldyn, 2010). Two microbiologically mediated mechanisms (Beutel 2006) are responsible for the release ammonia from anoxic lake sediments. The first mechanism is the inhibition of nitrification, the bacterial conversion of ammonia to nitrite then nitrate, under anoxic conditions with a resulting buildup of ammonia in sediment pore water, and its subsequent release into overlaying water (Rysgaard et al., 1994). The second mechanism is the depressed rates of ammonia assimilation by sediment microorganisms under anoxic conditions (Hansen and Blackburn, 1991). Since anaerobic microorganisms have much lower growth rates and cell yields than heterotrophic aerobic microorganisms, under anoxic conditions ammonia uptake and recycling decreases, resulting in sediment build up and release of ammonia (Patrick et al., 1996). Submerged aquatic plants may be particularly vulnerable to high water nitrogen concentrations because not only the roots, but also the leaves, are encompassed by water. Leaves, unlike roots, cannot regulate nitrogen uptake, e.g. through the reduction of root hairs (Marschner 1995).

Metal mobility and metal (loid) fate in sediment and in wetlands is governed by a number of soil factors and processes; e.g., adsorption/desorption reactions, precipitation/dissolution and complexation /decomplexation, salinity, organic matter content, sulphur (S) and carbonates, plant growth, pH and redox potential (E_H) as well as microorganisms activity (Du Laing et al., 2009; María-Cervantes et al., 2010).

Submerged aquatic plants are adapted to detoxify reduced elements by releasing root oxygen to the rhizosphre, but this phenomenon is also governed by Eh condition and microbial oxygen demand (Laskov et al., 2006). Normally, plants in wetland habitats will be defended against the uptake of these ions by oxidizing them at the root surface from surplus supplies of oxygen diffusing downwards from the shoot, or in the case of some rhizomatous species transported actively by humidity-induced pressure flow. When the supply of oxygen is interrupted, then these defense mechanisms fail and poisoning from toxic metal ions takes place. Excessive absorption of iron under flooded conditions can subsequently lead to tissue damage through enhanced active radical production. The concentration of metals in plants can be more than 100,000 times greater than in the associated water. Recent studies on HM toxicity revealed that these metals may cause molecular damage to plant cells either directly or indirectly through the excess generation of reactive oxygen species (ROS). These can damage membrane and inactive several enzymes by reacting very rapidly with DNA, lipids, pigments and proteins (Weckx and Clijsters, 1996). Thus low redox conditions exert substantial influence on physiological processes of plants.

Morphological Adaptations

Submerged macrophytes exposed to anoxic sediments have morphological and physiological means to tolerate anoxic conditions (Pedersen et al., 2004). The mechanisms and degree of anoxia tolerance of submerged macrophytes are species-specific. Wetland aquatic plants have developed adaptive structural and metabolic features to combat oxygen deficiency (Blokhina et al., 2003). Increased porosity enables the plants to perform longitudinal transport from aerial and photosynthetic oxygen to the rhizosphere. The internal aeration system of flood-in-olerant species is not sufficient to sustain an aerobic state in the root-soil environment. Another mechanism plants use to survive flooding is to change their timing of reproduction (Bloom et al., 1990). Most aquatic plants do not need cuticles or have thin cuticles (*Z. marina*), their stomata always open for they do not need to retain water, on each side of their leaves are a number of stomata, the presence of air sacs enables them to float, their roots are smaller light and feathery so they do not need to prop up the plants and roots are specialized to take in oxygen (Sand- Jensen et al., 2005). Shoot elongation of submerged macrophytes (*P. distinctus*) to escape from low oxygen medium (Harada et al., 2005), formation of aerenchyma (Seago et al., 2005), releasing radial oxygen loss (Wigand et al., 2000). Another common response of wetland plants to growth in strict anoxic soils is lignification and suberization of the root cell and an increase in air space particularly in the root cortex associated with development of new roots. Other aquatic plants often have diffusive root barriers to prevent radial O_2 loss or produce barriers to maintain sufficient O_2 transport to the root tips when sediments become reduced (Colmer, 2003).

Cellular and Antioxidative Response

Under reduced soil condition aquatic plants specially submerged macrophytes not only encounter with zero oxygen medium but also experiencing a high demand demand for their internal oxygen to carry on metabolic processes. Brix and Sorrell (1996) stated that the acceleration of oxygen loss by high external demand is a much more potent cause of tissue anoxia *in situ* than the mere absence of external oxygen. The oxygen limitation in the sediment cause reduction of the photosynthesis process of plants, thus induce in biomass declination (Armstrong and Drew, 2002). Regulation of anoxic metabolism is complex and not all the features are well established yet. Anaerobic carbohydrate catabolism provides some ATP during anoxia, albeit only 3–35% of the rate of energy production in aerobic cells (Gibbs and Greenway, 2003), but takes a higher toll on carbohydrates. Anoxic stress is always accompanied to some extent by oxidative stress (generation of ROS) and its consequences. The metabolic changes induced by anoxia (a drop of cytosolic pH, a decrease in adenylate energy charge, membrane lipid peroxidation, excess of NADH) may alter the antioxidant status of the tissue (Blokhina et al., 2003). Metabolic adaptation mainly includes avoidance of the energy starvation, prevention of anaerobic intermediate and end products. All these metabolic adaptations mutually allow tolerant plants to generate sufficient amount of energy, maintain mineral uptake and even to grow in total anoxia.

During O_2 deprivation glycolytic fermentation may to some extent be able to maintain cell metabolic processes at an adequate level (Gibbs and Greenway, 2003) but anaerobic metabolism cannot sustain active ion uptake, and after only a few minutes of cell anoxia, the ability of roots to take up nutrients is severely depressed (Drew, 1990).In previous studies (Barko and Smart, 1986; Jampeetong and Brix 2009, Zaman and Asaeda 2013) it was experienced that growth reduction of different macrophyte under anoxic condition.

Generation of ROS under low oxygen in crop plants and wetland plants has been detected by several researchers (see for example, Chirkova et al., 1998; Blokhina et al., 2001). High ammonium levels (NH_4-N $\geq$5 mg/L) caused significant acute biochemical changes in *P. crispus*, which potentially could lead to significant biochemical damage. The mechanisms include the formation of ROS induced by NH_4^+ during its metabolism by the plants (Nimptsch and Pflugmacher, 2007) and the imbalance of C–N reserves in plants stemming from the incorporation of NH_4^+ into free amino acids (FAA) at the expense of soluble carbohydrates (SC) consumption. On the other hand accumulation of nitrogen (N) in plant would result in more ROS generating by accelerating N metabolism in the plant (Foyer et al., 1994). SOD activity has negative correlation with ROS which ultimately effect on plant detoxification mechanism.

Ethanol that has accumulated under anoxia can then reacts with hydrogen peroxide and provide a rapid generation of acetaldehyde. Although ethylene is not toxic to plants it can induce potentially maladaptive changes in plant form. The presence of toxic elements had great affinity to bind with sulphydryl groups of enzymes and hence suppress the functioning of essential biological components (Van Assche, 1990).

Lipid and membrane integrity during oxygen deprivation are among the key factors in the survival of plants. Under anoxia a decrease in membrane integrity is a symptom of injury and is measured with different biochemical contents, among them activation of lipid peroxidation and enhanced electrolyte leakage (Chirkova et al., 1998) are two promising tools.

Plants under reduced environment were found more venerable to metal toxicity, by lowering its protein synthesis, as more than one metal was present at toxic level in the reduced experimental conditions (Monferrán et al., 2012, Zaman and Asaeda, 2013).

Conclusion

Plant colonization in metal polluted soils is hindered since the physicochemical parameters of these sites (e.g. extreme pHs, phytotoxic metal (oid) and depleted soil oxygen concentrations) are not favourable for most of the submerged species. High turbulence velocity on the other hand, acts as a disturbance factor for completely submerged macrophytes. Plant is exposed to the above mentioned stress conditions for a considerable period, might loss the scavenging ability of excess generation of ROS, in the long run. At the same time it effects on hormonal as well as antioxidant activities of plant. Macrophytes being an important component of fluvial ecosystems can reduce hydrodynamic energy from currents and waves, including the deposit of fine sediment and the anoxification of sediments. Thus it is important to provide suitable environment to the proper growth, reproduction, survival and colonization of aquatic macrophytes.

References

Armstrong J, Drew MC (2002) Root Growth and Metabolism under Oxygen Deficiency. *In* Plant Roots: The Hidden Half, Waisel Y, Eshel A, Kafkaf U, eds, Marcel Dekker, New York, USA,pp 729-761.

Asada K, Takahashi M (1987) Production and scavenging of active oxygen in chloroplasts. *In* DJ Kyle, CB Osmond, CJ Arntzen, eds, *Photoinhibition. Elsevier,* Amsterdam, pp 227–287.

Asaeda T, Fujino T, Manatunge J (2005) Morphological adaptations of emergent plants to water flow: a case study with *Typha angustifolia*, *Zizania latifolia* and *Phragmites australis*. *Freshwater Biol.* 50:1991–2001.

Asaeda T, Rajapakse L, Kanoh M (2010) Fine element retention as affected by annual collapse: *Sparganium erectum* as an ecosystem engineer in a lowland stream. *River Res. Appl.* 26:1153–1169.

Bal KD, Tjeerd JB, Kerst B, Eric S, Jonas S, Backx H, Meire P (2011) Trade-off between drag reduction and light interception of macrophytes: comparing five aquatic plants with contrasting morphology. *Funct. Ecol.* 25: 1197–1205.

Bari R, Jones J (2009) Role of plant hormones in plant defence responses. *Plant. Mol. Biol.* 69: 473–488.

Barko JW, Smart RM (1986) Sediment-Related Mechanisms of Growth Limitation in Submersed Macrophytes. *Ecology* 67: 1328-1340.

Beutel MW (2006) Inhibition of ammonia release from anoxic profundal sediments in lakes using hypolimnetic oxygenation. *Ecol. Eng*. 28: 271–279.

Blokhina OB, Virolainen E, Fagerstedt KV (2003) Antioxidants, Oxidative Damage and Oxygen Deprivation Stress: a Review. *Ann. Bot*. 91: 179-194.

Blokhina OB, Chirkova TV, Fagerstedt KV (2001) Anoxic stress leads to hydrogen peroxide formation in plant cells. *J. Exp. Bot*. 52:1179-1190.

Blom CWPM, Bögemann GM, Laan P, van der Sman AJM, van de Steeg HM, Voesenek LACJ (1990) Adaptations to flooding in plants from river areas. *Aquat. Bot*. 38: 29–47.

Boeger MRT, Poulson ME (2003) Morphological adaptations and photosynthetic rates of amphibious *Veronica anagallis-aquatica* L. (Scrophulariaceae) under different flow regimes. *Aquat. Bot*. 75: 123-135.

Brix H, Sorrell BK (1996) Oxygen stress in Wetland plants: Comparison of de-oxygenated and reducing root environments. *Funct. Ecol*. 10: 521-526.

Chapin III FS, Bloom AJ, Field CB, Waring RH. 1987. Plant responses to multiple environmental factors. *Bioscience* 37: 49–57.

Chehab W, Eich E, Braam J (2009) Thigmomorphogenesis: a complex plant response to mechano-stimulation. *J. Exp. Bot*. 60: 43–56.

Chirkova TV(1997) Cell membrane and plant resistance to stresses. *Russ. J. Plant. Physl*. 9: 12-17.

Colmer TD (2003) Long-distance transport of gases in plants: a perspective on internal aeration and radial oxygen loss from roots. *Plant. Cell. Environ* 26: 17–36.

Crawford RMM (1992) Oxygen availability as an ecological limit to plant distribution. *Adv. Ecol. Res*. 23: 93-285.

Crossley MN, Dennison WC, Williams RR, Wearing AH (2002) The interaction of water flow and nutrients on aquatic plant growth. *Hydrobiologia* 489:63–70.

Dat J, Vandenabeele S, Vranova´ E, Van Montagu M, Inze´ D, van Breusegem F (2000) Dual action of the active oxygen species during plant stress responses. *Cell. Mol. Life Sci*. 57:779–795.

DeLaune RD, Pezeshki SR, Pardue JH (1990) An oxidation–reduction buffer for evaluating the physiological responses of plants to root oxygen stress. Env*iron Exp. Bot*. 30: 243–247.

Devai I, DeLaune RD (1995) Formation of volatile sulfur compounds in salt marsh sediment as influenced by soil redox condition. *Org. Geochem*. 23: 289-287.

Downing-Kunz MA, Stacey MT (2012) Observations of mean and turbulent flow structure in a free-floating macrophyte root canopy. *Limnol. Oceanogr* 2: 62-79.

Doyle RD (2001) Effects of waves on the early growth of *Vallisneria americana. Freshwater Biol*. 46(3):389–397.

Drew MC (1990) Sensing soil oxygen. *Plant. Cell. Environ*. 13: 681-693.

Drew MC (1997) Oxygen deficiency and root metabolism: Injury and acclimation under hypoxia and anoxia. *Ann. Rev. Plant. Physiol. Plant. Mol. Biol*. 48: 223-250.

Du Laing G, Rinklebe J, Vandecasteele B, Meers E, Tack FMG (2009) Trace metal behaviour in estuarine and riverine floodplain soils and sediments: A review. *Science of The Total Environment* 407: 3972–3985.

Ellawala C, Asaeda T, Kawamura K (2011a) Influence of flow turbulence on *Chara fibrosa*: growth, stress, and tissue carbon content. *J. Freshwater Ecol*. 26: 507–515.

Ellawala C, Asaeda T, Kawamura K (2011b) The effect of flow turbulence on plant growth and several growth regulators in *Egeria densa Planchon. Flora* 206: 1085–1091.

Ellawala C, Asaeda T, Kawamura K (2012) The effect of flow turbulence on growth, nutrient uptake and stable carbon and nitrogen isotope signatures in *Chara fibrosa Ann. Limnol.- Int. J. Lim.* 48: 349-354.

Fonseca MJ, Kenworth WJ (1987) Effects of current on photosynthesis and distribution of sea grasses. *Aquat. Bot.* 27: 59-78.

Foyer CH, Descourvie'res P, Kunert KJ (1994) Protection against oxygen radicals: an important defense mechanism studied in transgenic plants. *Plant. Cell. Environ* 17: 507-523.

Foyer CH, Noctor G (2005) Redox homeostis and antioxidant signaling: a metabolic interface between stress perception and physiological responses. *Plant. Cell.,* 17: 1866–1875.

Gibbs J, Greenway H, (2003) Mechanisms of anoxia tolerance in plants. I. Growth, survival and anaerobic catabolism. *Funct. Plant. Biol.* 30: 1–47.

Gill SS, Tuteja N (2010) Reactive oxygen species and antioxidant machinery in abiotic stress tolerance in crop plants. *Plant. Physiol. Bioch.* 48:909-930

Goldyn H (2010) Changes in plant species diversity of aquatic ecosystems in the agricultural landscape in west Poland in the last 30 years. *Biodivers Conserv.* 19: 61–80.

Hansen LS, Blackburn TH (1991) Aerobic and anaerobic mineralization in marine sediment microcosms. *Mar. Ecol. Prog. Ser.* 75: 283–291.

Harada T, Satoh S, Yoshioka T, Ishizawa K (2005) Expression of sucrose synthase genes involved in enhanced elongation of pondweed (*Potamogeton distinctus*) turions under anoxia. *Ann. Bot.* 96: 683–692.

Herbinger K, Tausz M, Wonisch A, Soja G, Sorger A, Grill D (2002) Complex interactive effects of drought and ozone stress on the antioxidant defence systems of two wheat cultivars. *Plant. Physiol. Biochem.* 40: 691–696.

Hiraga S, Sasaki K, Ito H, Ohashi Y, Matsui H (2001) A large family of class III plant peroxidases. *Plant. Cell Physiol.* 42: 462–468.

Imlay JA, Linn S (1988) DNA damage and oxygen radical toxicity. *Science* 240:1302–1309.

Jackson LJ, Kalff J, Ramussen JB (1993) Sediment pH and redox potential affect the bioavailability of Al, Cu, Fe, Mn and Zn to rooted aquatic macrophytes. *Can. J. Fish Aquat. Sci* 50: 143–148.

Jampeetong A, Brix H (2009) Nitrogen nutrition of *Salvinia natans*: Effect of inorganic nitrogen form on growth, morphology, nitrate reductase activity and uptake kinetics of ammonium and nitrate. *Aquat. Bot.* 90 : 67–73.

Ke X, Li W (2006) Hormonal correlates of seedling growth of two *Vallisneria* species grown at different current velocities. *Hydrobiologia* 556: 243–249.

Laskov C, Horn O, Hupfer M (2006) Environmental factors regulating the radial oxygen loss from roots of *Myriophyllum spicatum* and *Potamogeton crispus. Aquat. Bot.* 84: 333–40.

Libreros-Minotta C, Tipton P (1995) A colorimetric assay for cytokinin oxidase. *Anal. Biochem.* 231: 339–341.

Lissner J, Mendelssohn IA, Anastasiou C (2003) A method for cultivating plants under controlled redox intensities in hydroponics. *Aquat. Bot.* 76: 93–108.

Liszkay A, Kenk B, Schopfer P (2003) Evidence for the involvement of cell wall peroxidase in the generation of hydroxyl radicals mediating extension growth. *Planta* 217: 658–667.

Madsen JD, Chambers PA, James WF, Koch EW, Westlake DF (2001) The interaction between water movement, sediment dynamics and submersed macrophytes. *Hydrobiologia* 44:71–84.

Madsen TV, Enevoldsen HO, Jørgensen TB (1993) Effects of water velocity on photosynthesis and dark respiration in submerged stream macrophytes. *Plant. Cell Environ.* 16:317–322.

María-Cervantes A, Conesa HM, Gonzalez-Alcaraz MN, Álvarez-Rogel J (2010) Rhizosphere and flooding regime as key factors for the mobilisation of arsenic and potentitally harmful metals in basic, mining polluted salt marsh soils. *Appl. Geochem.* 25: 1722–1733.

Marschner H (1995) Mineral nutrition of higher plants, 2nd edn. Academic press, London.

Misra AN, Biswal UC (1980) Effect of phytohormones on the chlorophyll degradation during aging of chloroplasts in vivo and in vitro. *Protoplasma* 105: 1-8.

Mohan BS, Hosetti BB (1997) Potential phytotoxicity of lead and cadmium to *Lemna minor* grown in sewage stabilization ponds. *Environ. Pollut.* 2:233–238.

Møller IM, Jensen PE, Hansson A (2007) Oxidative Modifications to Cellular Components in Plants. *Annu. Rev. Plant. Biol.* 58: 459-481.

Monferrán MV , Pignata ML, Wunderlin DA (2012) Enhanced phytoextraction of chromium by the aquatic macrophyte *Potamogeton pusillus* in presence of copper *Environ. Pollut.* 161:15-22.

Moore BC, Chen PH, Funk WH, Yonge D (1996) A model for predicting lake sediment oxygen demand following hypolimnetic aeration. *Wat. Resour.* Bull. 32: 1–9.

Moustakas, M., T. Lanaras, L. Symeonidis and S. Karataglis (1994) Growth and some photosynthetic characteristics on field grown *Avena sativa* under copper and lead stress. *Photosynthetica* 30: 389-396.

Musgrave A, Jackson MB, Ling E (1972) Callitriche stem elongation is controlled by ethylene and gibberellin. *Nat. New. Biol.* 1972; 238:93–96.

Navrot N, Rouhier N, Gelhaye E, Jaquot JP (2007) Reactive oxygen species generation and antioxidant systems in plant mitochondria. *Physiol. Plant.* 129: 185–195.

Nepf HM (1999) Drag, turbulence, and diffusion in flow through emergent vegetation. *Water. Resour. Res.* 35: 479-489.

Nielsen HD, Nielsen SL, Madsen TV (2006) CO_2 uptake patterns depend on water current velocity and shoot morphology in submerged stream macrophytes. *Freshwater Biol.* 51: 1331–1340.

Nimptsch J, Pflugmacher S (2007) Ammonia triggers the promotion of oxidative stress in the aquatic macrophyte *Myriophyllum mattogrossense*. *Chemosphere* 66: 708-714.

Nishihara GN, Ackerman JD (2006) The effect of hydrodynamics on the mass transfer of dissolved inorganic carbon to the freshwater macrophyte, *Vallisneria americana*. *Limnol. Oceanogr.* 51: 2734–2745.

Patrick WH, Gambrell RP, Faulkner SP (1996) Redox measurements of soils. *In* Sparks DL, ed, Methods of soil analysis, Part 3: Chemical methods, Soil Science Society of America, Madison, WI, USA, pp 645-657.

Pedersen O, Binzer T, Borum J (2004) Sulphide intrusion in eelgrass (*Zostera marina* L.). *Plant. Cell. Environ.* 27: 595–602.

Pezeshki SR, DeLaune RD (2012) Soil Oxidation-Reduction in Wetlands and Its Impact on Plant Functioning. *Biology 1*: 196-221.

Potters G, Pasternak TP, Guisez Y, Plame KJ, Jansen MAK (2009) Different stresses, similar morphogenic responses: integrating a plethora of pathways. *Plant. Cell. Environ.* 32:158–169.

Pujol D, Colomer J, Serra T, Casamitjana X (2010) Effect of submerged aquatic vegetation on turbulence induced by an oscillating grid. *Cont. Shelf. Res.* 30: 1019–1029.

Rysgaard S, Risgaard-Petersen N, Sloth NP, Jensen K, Nielsen LP (1994) Oxygen regulation of nitrification and denitrification in sediments. *Limnol. Oceanogr* 39: 1643–1652.

Sabbatini MR, Murphy KJ, Irigoyen JH (1998) Vegetation-environment relationships in irrigation channel systems of southern Argentina. *Aquat. Bot.* 60: 119-133.

Sand-Jensen K (2003) Drag and reconfiguration of freshwater macrophytes. *Freshwater Biol.* 48: 271–283.

Sand-Jensen K, Pedersen ML (2008) Streamlining of plant patches in streams. *Freshwater Biol.* 53: 714–726.

Sand-Jensen K, Pedersen O, Binzer T, Borum J (2005) Contrasting oxygen dynamics in the freshwater isoetid *Lobelia dortmanna* and the marine seagrass *Zostera marina. Ann. Bot.* 96: 613–623.

Schutten J, Dainty J, Davy AJ (2004) Wave-induced hydraulic forces on submerged aquatic plants in shallow lakes. *Ann. Bot.* 93: 333–341.

Schutten J, Dainty J, Davy AJ (2005) Root anchorage and its significance for submerged plants in shallow lakes. *J. Ecol.* 93:556–571.

Schutzendubel A, Schwanz P, Teichmann T, Gross K, Langenfeld-Heyser R, Godbold DL, Polle A (2001) Cadmium-induced changes in antioxidative systems, hydrogen peroxide content and differentiation in Scots Pine roots. *Plant. Physiol.* 127: 887-898.

Seago JL Jr, Marsh LC, Stevens KJ, Soukup A, Votrubova´ O, Enstone DE (2005) A re-examination of the root cortex in wetland flowering plants with respect to aerenchyma. *Ann. Bot.* 96: 565–579.

Siegel BZ (1993) Plant peroxidases—an organismic perspective. *Plant. Growth Regul.* 12: 303–312.

Szmeja J, Galka A (2008) Phenotypic responses to water flow and wave exposure in aquatic plants. *Acta. Soc. Bota.* Pol. 77:59–65.

Van Assche F, Clijsters H (1990) Effects of metals on enzyme activity in plants. *Plant. Cell. Environ.* 13:195-206.

Weckx JEJ, Clijsters HMM (1996) Oxidative damage and defense mechanisms in primary leaves of *Phaseolus vulgaris* as a result of root assimilation of toxic amounts of copper. *Physiol. Plantarum* 96: 506–512.

Westlake DF (1967) Some effects of Low-Velocity on the Metabolism of Aquatic Plants. *J. Exp. Bot.* 18: 187-205.

Wigand C, Wehr J, Limburg K, Gorham B, Longergan S, Findlay S (2000) Effects of *Vallisneria americana* (L.) on community structure and ecosystem function in lake mesocosms. *Hydrobiologia* 418: 137-146.

Wise RR, Naylor AW (1987) Chilling-enhanced photooxidation. Evidence for the role of singlet oxygen and superoxide in the breakdown of pigments and endogenous antioxidants. *Plant. Physiol.* 83:278-282.

Yang XC, Wang BC, Liu YY, Duan CR, Dai CY (2002) Biological effects of *Actinidia chinensis* callus on mechanical vibration. *Colloid Surface* B 25: 197-203.

Zaman T, Asaeda T (2013) Effects of NH_4-N concentrations and gradient redox level on growth and allied biochemical parameters of *Elodea nuttallii* (Planch.). Flora in press.

Zhang H, Shao M, Qiao Z, Yuan S, Wang X, Hua S (2009) Effect of phytohormones on fiber initiation of cotton ovule. *Acta. Physiol. Plant*. 31: 979–986.

In: Peroxidases
Editors: Lucas Bogaert and Noah Coppens

ISBN: 978-1-62808-261-6

Chapter VI

Plant Responses to Stresses: Role of Ascorbate Peroxidase in the Antioxidant Protection

***I. M. Huseynova*[1]*, D. R. Aliyeva*[1] *and J. A. Aliyev*[1,2]**

[1] Department of Fundamental Problems of Biological Productivity, Institute of Botany, Azerbaijan National Academy of Sciences, Baku, Azerbaijan

[2] Department of Plant Physiology and Biotechnology, Research Institute of Crop Husbandry, Ministry of Agriculture of Azerbaijan Republic, Baku, Azerbaijan

Abstract

The change in activities and numerous molecular forms of antioxidant enzymes is one of the parameters of plant biological tolerance against constantly changing environmental conditions. A group of genes involved in the synthesis of ascorbate is used in genetic engineering for the improvement of plant tolerance to biotic and abiotic stresses. The level of H_2O_2 in plant cells is controlled by two enzymes: catalase and ascorbate peroxidase (APX), though the key role in decomposition of H_2O_2 belongs to APX. Ascorbate peroxidase converts H_2O_2 into H_2O and O_2. The APX activity and electrophoresis spectra of its isoenzyme content have been investigated in plants subjected to abiotic (soil drought) and biotic (virus infections) stresses. The analysis of electrophoretic spectra of four isoforms in wheat leaves at the beginning of drought (flowering phase) revealed four isoforms. The amounts of APX isoforms increased to 7 during the wax ripeness phase both in irrigated plants and plants subjected to drought. The long-term drought led to the increase in the intensity of high-molecular isoforms and emergence of three additional isoforms of the enzyme. In this instance, the APX activity level remained higher in drought tolerant wheat genotypes compared with the sensitive ones. At the same time, in both tolerant and sensitive varieties, enhanced APX activity compared with the control variants was observed at the end of drought, which is apparently related to the synthesis of new molecules of the enzyme (i.e. it is stipulated by the changes of genome expression under stress). The immune enzymatic method revealed the presence of CMV virus in *C. sativus* L. plants and TMV in *L. esculentum* Mill. plants. The obtained results show significant differences in APX activities in all virus infected leaves compared with controls. The effect of the viral infection caused approximately a 1.6-fold increase of the APX activity in *L. esculentum* Mill. and a 2.1-fold increase in *C. sativus* L. compared with corresponding controls. Pathogenesis was accompanied by an increase in the quantity of APX isoforms, which reached 5 in leaves of *L. esculentum* Mill. and by the formation of 2 new isoforms and intensification of the enzyme electrophoresis spectra in infected leaves of *C. sativus* L. compared with the control variants. Thus, an increase in ascorbate peroxidase activity and *de novo* synthesis of its numerous molecular forms under stress conditions testify its involvement in the antioxidant defense system.

Keywords: Ascorbate-glutathione cycle, ascorbate peroxidase, soil drought, pathogens, Cucumber mosaic virus, Tomato mosaic virus, wheat genotypes, vegetable crops

Introduction

A significant part of agricultural plants perishes annually in consequence of the abiotic and biotic stress effects. One of the central problems of the modern plant biology is the investigation of genetically controllable mechanisms of plant tolerance to different stresses. Plants possess common mechanisms in their physiological tolerance to drought, low temperature and salinization. The cell degradation caused by different stresses triggers a range of biological, physiological and structural changes in plant cells. These changes especially occur in the expression of numerous genes. The primary change of the total plant metabolism subjected to any stress is emergence of free radicals in cytoplasm and other cell components, i.e. emergence of the so called oxidative stress. Development of photooxidative processes in a plant organism is accompanied by an excessive accumulation of active oxygen forms (AOF) in cells. The content of AOF is under multilevel control of antioxidant system enzymes (Mitteler, 2002; Apel and Hirt 2004). One of the parameters of plant biological tolerance to constantly changing environmental conditions is the change of activities and numerous molecular forms of antioxidant enzymes. Superoxide dismutase, which declines superoxide concentrations, plays a principal role in elimination of AOF (Merzlyak, 1989), as well as catalase, peroxidase and enzymes, involved in the ascorbate-glutathione cycle that remove an excess of peroxides (Scandalios, 1993).

Two strategic approaches using genetic engineering techniques currently exist for improvement of the plant tolerance to abiotic and biotic stresses. The first approach is based on the introduction into genome and overexpression of a gene or groups of genes, determining plant tolerance to stress. Groups of genes involved in the ascorbate synthesis are used in the second approach. Ascorbate peroxidase deactivates H_2O_2 by converting it into H_2O and O_2. The H_2O_2 level in plant cells is controlled by two enzymes: catalase and ascorbate peroxidase (APX), although a key role in the destruction of H_2O_2 belongs to APX (Asada, 1992).

Ascorbate peroxidase (EC 1.11.1.11) is related to a peroxidase family that includes the microbial peroxidase, yeast cytochrome c peroxidase and prokaryotic catalase-peroxidase (Takeda et al., 2000; Shigeoka et al., 1980; Wilkinson et al., 2002).

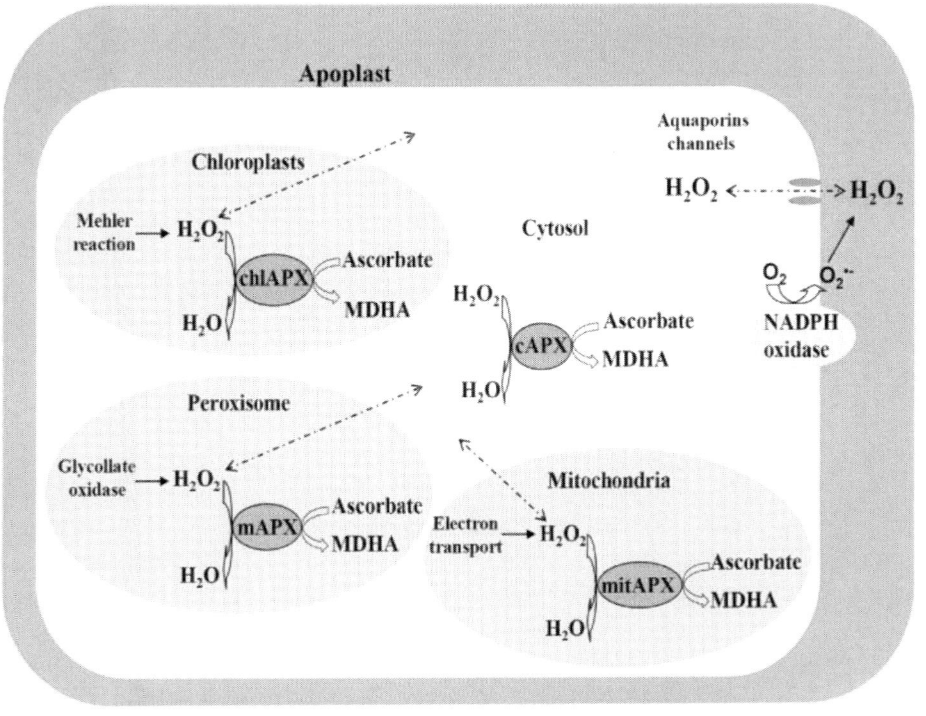

Figure 1. The ascorbate-glutathione cycle (*Halliwell- Asada cycle*). Enzymes of the ascorbate-glutathione cycle. The Mehler reaction-formation and accumulation of H_2O_2 in chloroplasts under stress conditions (Caverzan et al., 2012).

APX differs from guaiacol peroxidase not only in the specific donor, but also in the amino acid sequence and physiological functions. Ascorbate peroxidase (Mr 28-58 kDa) contains heme iron. It was detected in all intracellular compartments such as cytosol, chloroplasts, mitochondria and plant cell thylakoids (Mittler and Zilinskas, 1992; Bunkelman and Trelease, 1996; Ishikawa et al., 1996, 2008; Jimenes et al., 1998). Stromal and thylakoid-bound ascorbate peroxidases were isolated from chloroplasts. The thylakoid form is located predominantly adjacent to the acceptor part of PS I, providing the interception of H_2O_2 molecules close to their formation place.

The ascorbate peroxidase reaction is a central process of the whole cycle of reactions directed to the elimination of the main active oxygen forms in chloroplasts. The ascorbate-glutathione cycle is an example for such a cycle (Figure 1).

It functions in different compartments of cells: cytosol, peroxisomes, mitochondria, but mainly in chloroplasts of the plants lacking catalase, being the only means of H_2O_2 elimination. The cycle includes the following enzymes: ascorbate peroxidase, monodehydroascorbate peroxidase, dehydroascorbate peroxidase and glutathione reductase. During the functioning of the cycle, H_2O_2 is reduced in the presence of ascorbic acid. The oxidized glutathione is reduced in the presence of NADPH. The main enzyme of the cycle is ascorbate peroxidase. The enzyme is highly specific to ascorbate and rapidly loses activity in the absence of it. The ascorbate peroxidase reaction results in the formation of monodehydroascorbate-radical (MDA), which is reduced to ascorbate (Asc) by the stromal monodehydroascorbate reductase (EC 1.6.5.4):

$$H_2O_2 + 2Asc \rightarrow 2MDA + 2H_2O \quad (1)$$

$$2MDA + NAD(P)H{+}H^+ \rightarrow 2Asc + NAD(P)^+ \quad (2)$$

The thylakoid monodehydroascorbate reductase reacts to ferrodoxin instead of NAD(P)H. Disproportionation of monodehydroascorbate reductase leads to the formation of a relatively stable product dehydroascorbate (DHA), which is converted into ascorbic acid in the presence of dehydroascorbate reductase. At the same time oxidization of the reduced glutathione (GSH) occurs together with the formation of the appropriate dithiol (GSSG) (Foyer and Noctor, 2005):

$$2GSH + DHA \rightarrow Asc + GSSG \quad (3)$$

Photoreduction of O_2 in the electron-transport chain (ETC) of chloroplasts, or the Mehler reaction, is a well-known physiological alternative pathway of the electron transport in photosynthesis, including photosynthesis under abiotic stress conditions. However, there is still no consensus on the importance of the Mehler reaction. The question remains: does the activation of the photoreduction of O_2 promote the enhancement of the plant tolerance or lead to the initiation of the oxidative stress. In recent years extensive data have been obtained in a range of laboratories. Analysis of these data confirms the necessity of the formation and even accumulation of H_2O_2 in chloroplasts under stress conditions (Mubarashkina et al., 2006). The accumulation of H_2O_2 in chloroplasts has been shown to be the key factor of the systemic acquired acclimation of plants, in particular under conditions of high light intensity. It was observed that high H_2O_2 concentrations in chloroplasts could defend photosynthetic apparatus from photoinhibition. There are numerous reports on the photosystem I (PS I) as the main place of the formation of $O_2^{\bullet -}$ and H_2O_2 as a result of the electron transfer to O_2. At the same time according to the redox potentials of ETC carriers, the Mehler reaction is practically possible in any site of the ETC. Considering the necessity of the protonation of $O_2^{\bullet -}$, photoreduction of O_2 to H_2O_2 most probably occurs at the PS II level or the pool of plastoquinones.

Some modern investigations have confirmed this hypothesis proposed by scientists earlier. The study of the O_2 photoreduction in chloroplasts of plants, grown under different stress conditions and differing in their tolerance to unfavorable factors revealed the tendency to redistribution of electron flows in the ETC, namely, to decline in the O_2 photoreduction intensity at the PS I level and activation of the H_2O_2 photoformation in tolerant plants. Thus, the investigation of the photosynthetic electron transport in two different wheat varieties with contrasting drought tolerance showed the activation of O_2 photoreduction with the formation of H_2O_2 in the Mehler reaction in leaves of a tolerant variety and its decline in a sensitive variety under drought conditions. In contrast, the electron transport to O_2 with the formation of $O_2^{\bullet -}$ at the acceptor side of PS I decreased 7 times in chloroplasts of a tolerant variety and changed slightly in a sensitive variety as water deficiency of leaves increased (Allen and Hall, 1974).

Drought belongs to one of the most prevalent and crucial unfavorable environmental factors, which causes a sharp decline in the productivity of most agricultural crops (Abedi and Pakniyat, 2010; Passioura and Angus, 2010; Aliyev, 2012). Scientists of the Stanford University established that a 2°C increase in temperature above the average reduced the maturation period of wheat by nine days. This led to a 20% yield decline. Severe water deficiency in soil detains biosynthesis of organic compounds and intensifies hydrolysis leading to the disturbance of growth processes (Sayar et al., 2008; Tas and Tas, 2007; Khan and Naqvi, 2010). The process of oxidative phosphorylation and ATP synthesis is also disturbed. The activity of proteases increase leads to a decline in protein content and an increase in low-molecular products of their decomposition; the carbohydrate exchange is disturbed (Al-Khatib and Paulsen, 1999). Soil drought can also facilitate the generation of active oxygen forms (AOF), especially when it is accompanied by high solar insolation (Fu and Huang, 2001). Over the past decades, drought tolerance of plants has been successfully increased in many countries (Fischer et al., 2006, Burke et al., 2006). Plant tolerance to water deficiency can certainly be provided by different mechanisms, the efficiencies of which depend on climatic conditions (Lafitte et al., 2007; Takase et al., 2011). A clear relationship between drought tolerance and contents of osmotics has been revealed, and introduction of genes encoding enzymes, catalyzing the formation of osmotic active products into the plant genome, is being used over the last few years. For example, tolerance to increased temperatures improved when the peroxisomal APX gene is expressed in transgenic Arabidopsis plants (Shi et al., 2001). Transgenic tobacco plants with the expression of the *tAPX* gene in chloroplasts were characterized by an enhanced tolerance to freezing and high light intensities (Yabuta et al., 2002). Tobacco transformants expressing SOD cDNA of alfalfa appeared to be resistant to paraquat, heavy metals, hydrogen peroxide and sodium chloride and can grow back after 35-day periods of drought (Oberschall et al., 2000).

Anthropogenic impacts on biota have currently become one of the most important environmental factors. Therefore an assessment of the adaptive capacity of living organisms that inhabit in radiation biocenoses is particularly relevant. The effect of pollutants on the production of AOF has been extensively studied lately. Ions Fe^{2+}, Fe^{3+} and/or Cu^{2+} have been shown to catalyze the formation of free radicals in the Fenton and Haber-Weiss reactions (Dat et al., 2000; Mitteler, 2002). In the Fenton (equations 4, 5) and Haber-Weiss (equation 6), reactions with the involvement of the mentioned ions superoxide radical and hydrogen peroxide form the hydroxyl radical, which is the most powerful oxidizer (Dat et al., 2000; Mitteler, 2002):

$$O_2^{\bullet -} + Fe^{+3} (Cu^{+2}) = O_2 + Fe^{+2} (Cu^{+}) \quad (4)$$

$$Fe^{+2} (Cu^{+}) + H_2O_2 = Fe^{+3} (Cu^{+2}) + HO^{\bullet} + HO^{-} \quad (5)$$

$$O_2^{\bullet -} + H_2O_2 = HO^{\bullet} + HO^{-} + O_2 \quad (6)$$

In addition to the mentioned metals, zinc, nickel, aluminum, cadmium, lead, etc. also cause an active generation of AOF (Devi, Prasad, 2005; Fang, Kao, 2000). Cell compartmentation of this process is still unclear. Polesskaya (Polesskaya, 2007) suggested that the reason for the oxidative stress under these conditions could be the effect of the simultaneous toxic diffuse of metal ions on the most enzymatic systems and membranes, as well as declines in the efficiency of antioxidant systems as a consequence of the diversion of reduced glutathione to the synthesis of phytochelatins, which detoxify them as a result of conjugation with pollutants. Transgenic plants of *Pinus virginiana* Mill. with increased antioxidant enzyme (ascorbate peroxidase, glutathione reductase) activities were described (Tang et al., 2005). These plants were shown to possess an enhanced tolerance to different stresses, which increase their survivability, particularly in metal-polluted soils. To maintain homeostasis in the presence of heavy metal toxic concentrations plants synthesize phytochelatins from glutathione. They bind heavy metals and form complexes with Cd, Ag, Pb, Zn, Cu and Hg, providing their detoxification (Clemens et al., 1999). However, activities of the antioxidant enzymes of plants from nature ecosystems, living for a long time under conditions with increased but not lethal contents of such metals as nickel, zinc, iron and manganese have not been studied sufficiently. Radionuclides penetrating into plant tissues out of soil cause significant changes in cell metabolism, operating as stress factors, which is the reason for increased APX activity when soil is polluted with different natural radionuclides. Then plants proceed to stress state and their signal warning systems are activated. APX is one of the components of such systems. Therefore the increase in its activity in response to stressors, in this case - radionuclides - is quite clear. This enzyme could probably be considered as a marker of the plant stress state when soil is contaminated with different radionuclides.

In addition to the data on the intensive generation of active oxygen forms under the influence of abiotic factors, a large amount of experimental data on the induction of AOF under biotic stress have been compiled. The intensification of the generation of AOF causes damage to plants by bacteria, fungi and microorganisms (Kolupayev, 2007; Maksimov, Cherepanov, 2006; Mohamed, Raldugina et al., 2006; Tarchevskiy, 2002). In vegetables more than 100 specific viruses of different taxonomic groups that differ in their biological, economic, serological properties and modes of transmission were registered. Seven exciters, spread by contact, insects and through soil were identified only in cucumber. 36 viruses and their strains from 12 genera were registered in tomato (Keldish, 2006). Viral infections of crops cause significant yield losses, impairment of the quality of agricultural products and are considered as a serious threat for food security. This problem concerns all food, feed and industrial crops cultivated in any region of the Earth and is especially actual for vegetatively propagated plants, since progressive accumulation of viruses in a range of generation leads to the complete infection and degradation of a variety. Annual losses of one crop in a certain region caused by viruses are often expressed in hundreds of millions (and billions) of dollars (Kegler et al., 1986; Hull and Davies, 1992; Waterworth and Hadidi, 1998; Strange and Scott, 2005).

The important problem of modern biology is the study of mechanisms of the induction of plant resistance to the effects of pathogens. In response to the pathogen infection, defense reactions, such as enhancement of the barrier properties (lignification), expression of genes and synthesis of proteins, including enzymes and especially peroxidases are switched on (Rogozhin, 2010). The role of peroxidases in plant defense mechanisms during pathogenesis was extensively studied (Maksimov, Cherepanova, 2006; Minibayeva, Gordon, 2003; Tarchevskiy, 2002; Baker et al., 1997; Bolwell et al., 2002; Zhao, Sakai, 2003; Kuznaik, Sklodowska, 2005). It was established that some peroxidases were located in the apoplastic space of plant cells (Betwick et al., 1998). The peroxidase levels were shown to be higher in cells of plant varieties resistant to pathogens compared with the sensitive ones. Similar data was obtained particularly for potato (Graskova et al., 2005) and pea (Kozyavina, 2012).

Therefore the investigation of the protective role of ascorbate peroxidase in plant adaptation to the changing environmental conditions, including drought and pathogen microbiota, is highly relevant.

Materials and Methods

Six durum (*Triticum durum* Desf.) and bread (*Triticum aestivum* L.) wheat genotypes contrasting for drought tolerance, such as Barakatli-95, Gyrmyzy bugda, Azamatli-95 (drought tolerant), Garagylchyg-2, Gyrmyzy gul-1, Giymatli-2/17 (drought sensitive) were chosen as investigation objects. Plants were grown in the field of the Absheron Experimental Station of the Research Institute of Crop Husbandry (Baku, Azerbaijan) under conditions with a normal water supply and water deficiency (Figure 2). Control plants were grown under normal irrigation conditions. The use of just such a model system allowed us to study dynamics of the activities of antioxidant enzymes at the earliest stages of drought and till the end of the vegetation.

Figure 2. Wheat genotypes grown under conditions of severe water deficiency at the experimental station of the Research Institute of Crop Husbandry.

The objects of study were also samples of various vegetable crops grown in field conditions at the Experimental Station of the Research Institute of Vegetable Growing (Baku, Azerbaijan), including mainly pepper, tomato and eggplant with symptoms of viral diseases. At first phenological and phytopathological investigations were carried out. The first symptoms of the infection include deformation of young leaves, yellow engraving and enlightenment of veins. As the infection progressed symptoms of chlorosis, mosaic, mottling and distortion of leaves were observed (Figure 3). An *immunoenzyme analysis*-based *test system* revealed the presence of viruses CMV (Cucumber mosaic virus) in *Cucumis sativus* L. and TMV (Tomato mosaic virus) in *Lycopersicon esculentum* Mill. (Clark, Adams, 1977).

Cucumis sativus L. — *Piper longum* L. — *Solanum melongena* L. — *Lycopersicon esculentum* Mill.

Figure 3. Distortion, chlorosis, mosaic, yellow gravity and enlightenment of veins in young leaves of viral infected vegetable crops.

Activities and molecular forms of APX were studied in wheat genotypes subjected to drought and vegetable crops infected with viruses. A sample of plant tissue was ground in 50 mM K-phosphate buffer (pH 7.8) containing 1 mM PMSF, 1 mM ascorbic acid and 1% polyvinylpyrrolidone. The homogenate was filtered through 4 layers of cheesecloth and then centrifuged at 15,000xg at 4°C for 15 min. The obtained supernatant was used for the analysis. The activity of ascorbate peroxidase (EC 1.11.1.11) was determined according to (Verma and Dubey, 2003). The reaction mixture contained 50 mM K-phosphate buffer (pH 7.0), 0.2 mM ascorbic acid, 0.2 mM EDTA, and the enzyme extraction. The reaction was initiated by adding 20 mM hydrogen peroxide. The optical density was measured at 290 nm for 120 s. with a spectrophotometer ULTROSPEC 3300 PRO ("AMERSHAM", USA). The enzyme quantity changing the optical density of the reaction mixture by 0.001 units per unit time was assumed as the enzyme activity unit.

Qualitative changes of the enzyme activity were investigated with native polyacrylamide gel (PAGE) using the Laemmli method (Laemmli, 1970) of gel electrophoresis with some modifications. An equal amount of the enzyme extract was mixed with bromophenol blue and glycerol to a final concentration of 12.5% (v/v). Electrophoresis in 10% PAAG with a thickness of 0.75 mm and length of 16 cm was performed at 4°C for 3 hours at a steady current of 30 mA. 2 mM sodium ascorbate was added to the electrode buffer to detect APX activity. For visualization of ascorbate peroxidase, the gel was incubated in a solution containing 50 mM potassium phosphate buffer (pH 7.0) and 2 mM Na-ascorbate for 30 min. Then the gel was incubated in a solution containing 50 mM potassium - phosphate buffer (pH 7.0), 4 mM sodium ascorbate and 2 mM H_2O_2, for 20 min and stained in a solution containing 50 mM potassium phosphate buffer (pH 7.8), 28 mM TEMED and 2.45 mM nitroblue tetrazolium, for 15 min by mixing with a shaker (Mittler and Zilinskas, 1993). The protein amount was determined by the method of Sedmak (Sedmak and Grossberg, 1977). Bovine serum albumin was used for the construction of standard curve.

Statistical analysis. Data of the determination of the ascorbate peroxidase activity are presented as arithmetic means and standard deviations of two separate experiments carried out with three biological replicates.

Results and Discussion

Ascorbate peroxidase is one of the key components of the cell protection system against harmful effects of H_2O_2. The role of ascorbate peroxidase is very important in the oxidative process. Its activity and izoenzyme content strongly depend on the effects of both biotic and abiotic external factors. The enzyme is sensitive to changes in the composition of soils to pollution and salinization, the impact of viral and bacterial pathogens. The APX activity also increased as metabolism intensified - during the spring active growth and flowering period (when generative organs and fruit formed) (Rogozhin, 2004). The study showed that the magnitude of the constitutive activity of ascorbate peroxidase was high in wheat leaves at the beginning of the experiment (i.e. flowering stage) (Figure 4). The enzyme activity decreased in the phase of milk ripeness; afterwards it increased in the phase of wax ripeness in drought tolerant and decreased in drought sensitive genotypes. The enzyme activity decrease can be explained by the rapid inactivation of its pool during the catalytic reaction. Resynthesis of new enzyme molecules (or isozymes), apparently provided the activity increase in the wax ripeness phase. The synthesis of new molecules could be initiated by the formed peroxide considering its signaling role (Garifzyanov et al., 2011).

Dynamics of the APX functioning differed in irrigated plants of both tetraploid and hexaploid varieties of wheat. The maximum APX activity in tetraploids was observed at the end of the flowering phase, while in hexaploids it was detected at the end of the earring phase. The most maximum APX activity among all studied genotypes was found in drought tolerant genotypes Azamatli-95 and Gyrmyzy bugda at the end of the flowering stage (1.51 and 1.49 μmol/mg min, respectively).

(A)

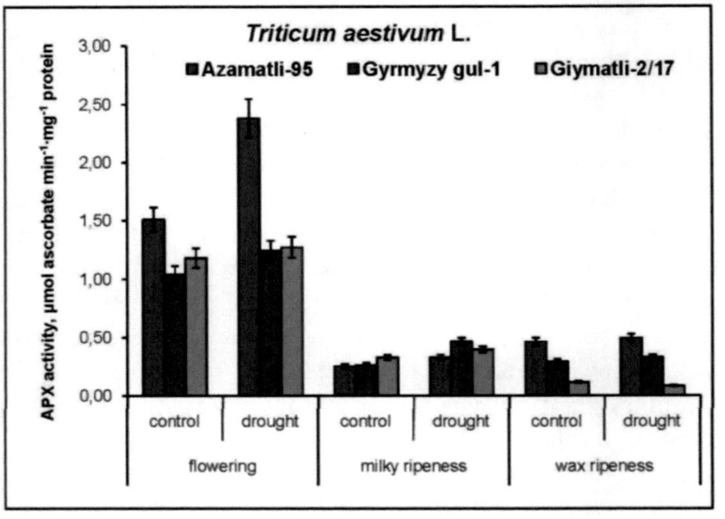

(B)

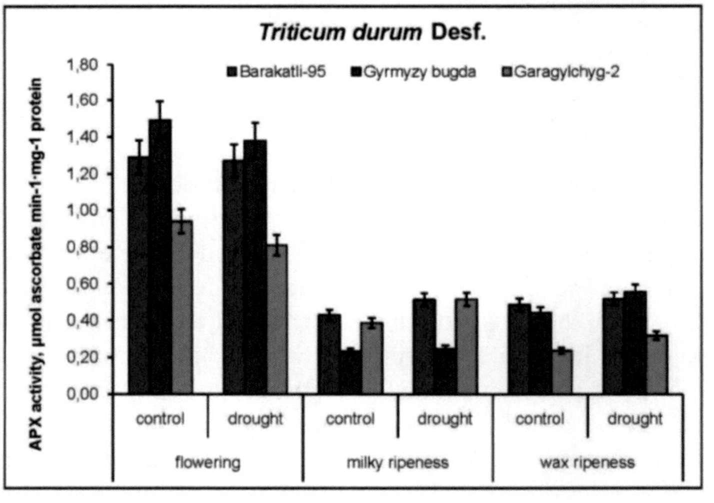

Figure 4. Dynamics of the ascorbate peroxidase activity in bread (A) and durum (B) wheat genotypes subjected to long-term soil drought.

The enzyme activities in control and drought stressed variants of tetraploid varieties did not differ significantly at the end of the flowering phase. In contrast, in hexaploids subjected to stress, the enzyme activity increased significantly at the end of the flowering phase compared with the control. However, in drought sensitive Giymatli-2/17 the enzyme activity was higher compared with the control variants during the whole ontogenesis (except for the wax ripeness phase). Thus, the reaction of the systems providing defense against harmful effects of water stress in investigated varieties of wheat were different. The level of the enzyme activity appeared to depend on the varieties of wheat, duration of drought and stages of leaf development.

The electrophoretic separation of numerous molecular forms of ascorbate peroxidase revealed significant qualitative (presence of additional bands in electrophoregram) and quantitative (changes of the color intensity of the individual bands) changes in expressivity of individual isoforms of APX during the soil drought (Figure 5A, B).

Gene expression is a rapid response to the influence of stress factors. The electrophoretic separation showed that the amount of APX isoforms varied in leaves depending on the growth stages of the plants. The least amount was observed at the beginning of the flowering phase (4 isoforms), the most - at the end of the wax ripeness phase (7 isoforms). During the drought, the amounts of APX isoforms reached seven in both variants (irrigated and drought-stressed). The increase in the intensity of the high-molecular isoform and appearance of 3 additional APX isoforms were observed in the wax ripeness phase (Figure 5B).

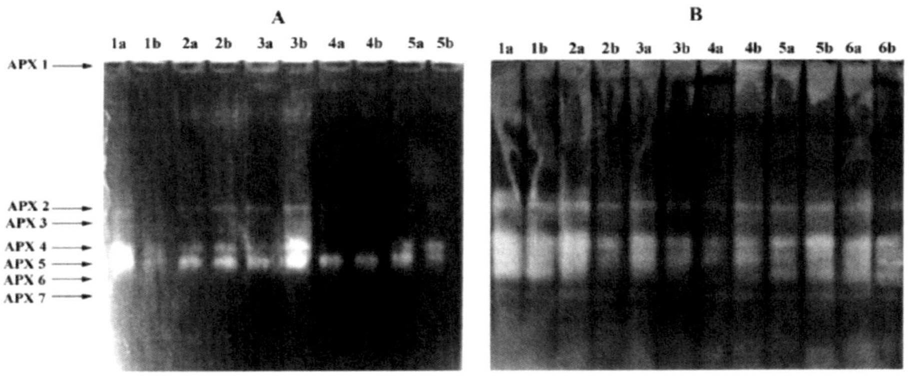

Figure 5. Electrophoretic spectra of ascorbate peroxidase in leaves of wheat grown under drought conditions. A - flowering phase, B - wax ripening phase, 1 - Garagylchyg-2, 2 - Gyrmyzy gul-1, 3 - Azamatli-95, 4 - Giymatli-2/17, 5 - Barakatli-95, 6 - Gyrmyzy bugda: a - irrigated, b - subjected to drought; Electrophoresis was carried out in 10% PAAG using tris-glycine buffer, pH 8.3 (with adding 2mM sodium ascorbate) at 4°C for 3 hours at 30mA stable current. 35μg of protein samples were applied to a gel lane.

It should be noted that the constitutive level of the APX activity was higher in tolerant genotypes, Azamatli-95 and Barakatli-95, than in sensitive ones. Moreover, at the end of the drought, increases in the APX activity compared with the controls were observed both in tolerant and sensitive varieties (except for Giymatli-2/17), which is apparently related to the synthesis of new enzyme molecules (i.e. due to the change in genome expression under stress). The enzyme activity increase under stress could be due to the activation of its latent forms or/and synthesis of new enzyme molecules.

Thus, the simultaneous increase in the activity of APX and the corresponding proteins were observed under salt stress in chloroplasts of pea (Gomez et al., 2004), and wheat (Navari-Izzo et al., 1998), in leaves of a tolerant variety of tomato *Lycopersicon penneli* (Mittova et al., 2003), confirming an intensification of the enzyme synthesis under the influence of stress factors.

The increase in the APX activity, which we observed in wheat leaves, could be related to an enhancement of H_2O_2 concentrations. Adapted wheat plants were shown to be characterized by enhanced activities of antioxidant enzymes (Selate and Khanna-Chopra, 2006). In contrast, activities of ascorbate peroxidase, superoxide dismutase and enzymes of the ascorbate-peroxidase cycle decreased, causing accumulation of H_2O_2 excessive levels in cells of unadapted plants under drought. This caused a dysfunction of chloroplasts and mitochondria due to damage to their membranes, and, therefore, all metabolic processes in cells and the growth of plants in general.

It was recently found that the activation in expression of an ascorbate peroxidase gene *APX2* in leaves of Arabidopsis happened in the presence of the extracellular H_2O_2 pool (Bechtold et al., 2008). A treatment of cultivable soybean cells with exogenous H_2O_2 also led to changes in transcription levels of *cAPX* (Kausar et al., 2012).

These results confirm that the expression of *cAPX* genes are regulated at the transcriptional level and activated in response to H_2O_2 in cells. The accumulation of H_2O_2 through redox-signaling can play an important role in the regulation of expression of *cAPX* genes under stress conditions. The increased activity of ascorbate peroxidase confirms its apparent participation in the antioxidant system under drought conditions. Therefore, an intensification of the gene expression of antioxidant enzymes catalase (Choi et al., 2008) and ascorbate peroxidase (Park et al., 2008) by means of genetic engineering methods increased tolerance of rice plants to drought. The activities of ascorbate peroxidase in all investigated samples of infected leaves differed significantly from that in control variants (Figure 6). Effects of the viral infection caused approximately a 1.6-fold increase in the enzyme activity in studied samples of *Lycopersicon esculentum* Mill. reaching 0.36 μmol/mg min. As shown in figure 6, slight changes occurred in the APX activity in the drought stressed *Solanum melongena* L. and *Piper longum* L. compared with the control variants. Thus, in infected samples of *Solanum melongena* L. and *Piper longum* L. the enzyme activity increased only by 24% and 26% compared with the controls and was estimated to be 0.34 and 0.44 μmol/mg min, respectively.

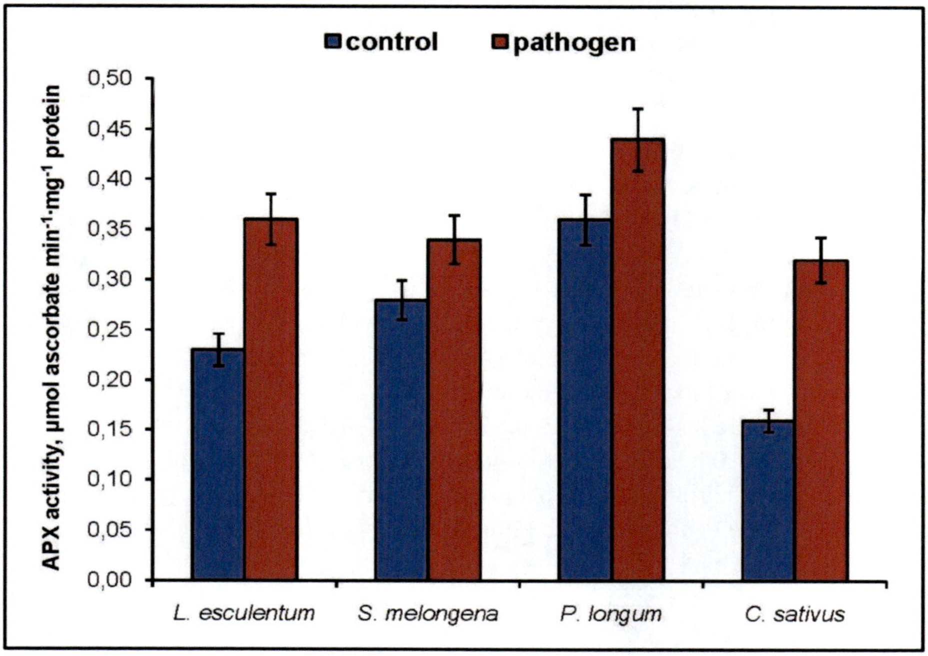

Figure 6. Activity of ascorbate peroxidase in leaves of vegetable crops infected with the viruses.

Table 1. Changes in isoform amounts in leaves of the vegetable crops infected with viruses

Plants	Isoform amounts in control plants	Isoform amounts in infected plants	Changes in isoform amounts after the infection
Lycopersicon esculentum	3	5	+2
Mill.	3	4	+1
Solanum melongena L.	3	4	+1
Piper longum L.	3	3	-
Cucumis sativus L.			

"+" - increase in the number of isoforms; "–" - the number does not change.

According to figure 6, the APX activity in infected *Cucumis sativus* L samples was enhanced significantly. In these samples the activity increased 2.1 times, reaching 0.32 μmol/mg min. Varieties of crops, as shown in the majority of reports, are characterized by a higher initial level of the antioxidative activity or are able to increase this activity more rapidly and they are more tolerant to the oxidative damage under stress conditions (Zhang, Kirkham 1996; Kang, Salveit 2002).

Pathogens infecting plants cause a rapid activation of peroxidase, often accompanying the emergence of numerous new forms and vanishing of the other forms of this enzyme (Nagy et al., 2004; Mayer et al., 2001). As shown in table 1, new APX forms did not appear in infected *Cucumis sativus* L. plants. In this case, 3 isoforms of APX were detected. The observed intensification of lane colors in the gels confirms the increase in the enzyme activity in this plant after the infection with pathogens. In the other investigated plants the infection led to the increase in the amounts of isoforms.

The intensity and composition of the electrophoretic spectrum of APX in leaves of the infected *Lycopersicon esculentum* Mill. changed compared with that of the normally developed plants. Thus, 3 isoforms were observed in the electrophoresis of the control variant. Pathogenesis under the influence of the viral infection was accompanied by the increase in amounts of APX isoforms to five and the formation of two new high-molecular forms. According to the results of the analysis during the pathogenesis the constitutive level of the enzyme activity in *Solanum melongena* L. leaves was higher compared with the control variant. Three low-molecular forms of APX occur in healthy plants, while four isoforms and an additional intensive high-molecular isoform were detected in plants with viral infections. The constitutive level of the APX activity in *Piper longum* L. also changed in infected plants, since three isoforms and an additional high-molecular form were observed in the stressed variant. The obtained data suggest that in plant cells exposed to the pathogenesis of the viral infection, the accumulation of peroxides at the early stages of the infection was inhibited by the increase in the activity of peroxidases.

Conclusion

The enhanced activity of ascorbate peroxidase and *de novo* synthesis of its various molecular forms at different stresses, testify to its undoubted involvement in the antioxidant defense system. Using APX as a marker of the stress state allows better characterization of the protective mechanisms of plants and finding approaches for the diagnostics of the tolerance to stress factors in different agricultural crops. The obtained data can be used to develop the strategy and tactics of breeding programs to create varieties tolerant to abiotic and biotic stressors.

Acknowledgments

This work was supported by a grant (EİF -2010-1(1)-40/24-M-20) of the Science Development Foundation under the President of the Republic of Azerbaijan.

References

Abedi T., Pakniyat H. (2010). Antioxidant enzyme changes in response to drought stress in ten cultivars of Oil seed rape (*Brassica napus* L.). *Czech J. Genet. Plant Breed.*, v.46(1), p.27-34.

Aliyev J.A. (2012). Physiological and molecular bases of drought tolerance in wheat (*Triticum* L.) genotypes. In: D.F. Neves, J.D. Sanz (eds.) *Environmental Science, Engineering and Technology. Drought: New Research*, Nova Science Publishers, Inc., New York, v.2, p.47-95.

Al-Khatib K. and Pauls G.N. (1999). High-temperature effects on photosynthetic processes in temperate and tropical cereals. *Crop Si. Soc. Amer.,* v.39, p.119-125.

Allen J.F. and Hall D.O. (1974). The relationship of oxygen uptake to electron transport in photosystem I of isolated chloroplast: the role of superoxide and ascorbate. *Biochem. Biophys. Res. Commun.*, v.58, p.579-585.

Apel K. and Hirt H. (2004). Reactive oxygen species: Metabolism, oxidative stress, and signal transduction. *Annu Rev Plant Biol.,* v.55, p.373-399.

Asada K. (1992). Ascorbate peroxidase – a hydrogen peroxide-scavenging enzyme in plants. *Physiol. Plantarum*, v.85, p.235-241.

Baker C.J. and Orlandi E.W. (1995). Active oxygen in plant/pathogen interactions. *Annual Review of Phytopathology*, v.33, p.299-321.

Bechtold U., Richard O., Zamboni A. et al. (2008). Impact of chloroplastic- and extracellular-sourced ROS on high light-responsive gene expression in *Arabidopsis. J Exp. Botany*, v.59, p.121-133.

Bolwell G.P., Bindschedler L.V., Blee K.A., Butt V.S., Davies D.R., Gardner S.L., Gerrish C., Minibayeva F. (2002). The apoplastic oxidative burst in response to biotic stress in plants: a tree component system. *J Exp Bot.*, v.53, p.1367–1376.

Bunkelman J., Trelease R. (1996). Ascorbate peroxidase: a prominent membrane protein in oilseed glyoxysomes. *Plant Physiol.*, v.110, p.589-598.

Burke E.J., Brown S.J., Christidis N. (2006). Modeling the recent evolution of global drought and projections for the twenty-first century with the Hadley Centre climate model. *J. Hydrometeorol.*, v.7, p.1113–1125.

Caverzan A., Passaia G., Barcellos Rosa S., Werner Ribeiro C., Lazzarotto F., Margis-Pinheiro M. (2012). Plant responses to stresses: Role of ascorbate peroxidase in the antioxidant protection. *Genetics and Mol Biol.*, v.35(4), p.1011-1019.

Choi H.J., Joo J.S., Song S.I. (2008). Overexpression of catalase gene increases of tolerance to drought stress in transgenic rice. *Physiol. Plant.*, v.133, p.5-19.

Clark M.F. and Adams A.N. (1977). Characteristics of the microplate method of enzyme-linked immunosorbent assay for the detection of plant viruses. *Journal of General Virology*, v.34, p.475-483.

Clemens S., Kim E.J., Neumann D., Schroeder J.I. (1999). Tolerance to toxic metals by gene family of phytochelatin synthases from plants and yeast. *EMBO J.*, v.18, p.3325–3333.

Dat J., Vandenabeele S., Vranova E., Van Montagu M., Inze D., Van Breusegem F. (2000). Dual action of the active oxygen species during plant stress responses. *Cellular and Molecular Life Sciences,* v.57, p.779-795.

Devi S.R., Prasad M.N.B. (2005). Antioxidant activity of Brassica juncea plants under the influence of high Cu concentrations. Plant physiology, v.52, p.233-238 (in Russian).

Fang W.C., Kao C.H. (2000). Enhanced peroxidase activity in rice leaves in response to excess iron, copper, and zinc. *Plant Science,* v.158, p.71-76.

Fischer G., Tubiello F.,Van Velhuizen H., Wiberg D. (2006). Climate change impacts on irrigation water requirements: effects of mitigation 1990–2989. *Technol. Forecasting Soc. Change,* v.74, p.1083–1107.

Foyer C.H. and Noctor G. (2005). Redox homeostasis and antioxidant signaling: A metabolic interface between stress perception and physiological responses. *Plant Cell,* v.17, p.1866-1875.

Fu J., Huang B. (2001). Involvement of antioxidants and lipid peroxidation in the adaptation of two cool-season grasses to localized drought stress. *Environm. and Experimen. Botany*, v.45, p.105-114.

Garifzyanov A.R., Zhukov N.N. (2011). The formation and physiological reactions of active oxygen forms in plant cells. Modern problems of science and education, №2; URL: *www.science-education.ru/96-4600* (in Russian).

Gomez J., Jimenez A., Olmos E., Sevilla F. (2004). Location and effects of long-term NaCl stress on superoxide dismutase and ascorbate peroxidase isoenzymes of pea (*Pisum sativum* cv. Puget) chloroplasts. *J. Exp. Botany*, v.55, p.119-130.

Graskova I.A., Antipina I.V., Potapenko O. Y., Voynikov V.K. (2005). Pathogen influence on the dynamics of the extracellular peroxidase activity in potato suspension cells. *J. Stress Physiology and Biochemistry*, v.1, p.16-20 (in Russian).

Hull R. and Davies J.W. (1992). Approaches to non-conventional control of plant virus diseases. *Crit. Rev. Plant Sci.*, v.11, p.17-33.

Ishikawa T. and Shigeoka S. (2008). Recent advances to ascorbate biosynthesis and the physiological significance of ascorbate peroxidase in photosynthesizing organisms. *Biosci Biotechnol Biochem.*, v.72, p.1143-1154.

Ishikawa T., Sakai K., Yoshimura K., Takeda T. and Shigeoka S. (1996). cDNA encoding spinach stromal and thylakoid bound ascorbate peroxidase, differing in the presence or absence of their 3'-coding regions. *FEBS Lett* 384, 289-293.

Ishikawa T., Yoshimura K., Sakai K. et al. (1998). Molecular characterization and physiological role of a glyxisome-bound ascorbate peroxidase from spinach. *Plant Cell Physiol.*, v.39, p.23-34.

Jimenez A., Hernandez J.A., del Rio L.A., Sevilla F. (1997). Evidence for the presence of the ascorbate-glutathione cycle in mitochondria and peroxisomes of pea leaves. *Plant Physiol.*, v.114, p.275-284.

Kang H.M., Saltveit M.E. (2002). Enzymatic antioxidant defense systems in chilled and heat shocked cucumber seedlings radicles. *Physiologia Plantarum*, v.115(4), p.571-576.

Kausar R., Hossain Z., Makino T., Komatsu S. (2012). Characterization of ascorbate peroxidase in soybean under flooding and drought stresses. *Mol Biol. Reports*, v.39(2), p.10573-10579.

Kegler H., Klaynhempel H., Ertel K. (1986). Yield losses at viral diseases. *Control of plant viral diseases*. M. (in Russian).

Keldish M.A. (2006). Plant Protection and Quarantine, v.5, p.24-25 (in Russian).

Khan N., Naqvi F.N. (2010). Effect of water stress on lipid peroxidation and antioxidant enzymes in local bread wheat hexaploids. *Journal of Food, Agriculture and Environment*, v.8 (2), p.521-526.

Kolupayev Y.E. *Active oxygen forms in plants exposed to the effect of stressors: formation and possible functions*. Herald of the Kharkov National agricultural university (ser. biol.), v.3 (12), p.6-26 (in Russian).

Kozyavina K.N. (2010). *The change in activity of the antioxidant enzyme system and fragmentation of DNA of Pisum sativum L. in apoptosis and necrosis under the influence of 2 valent ions of magnesium and extract of trichoderma.* PhD thesis, Voronezh, 24pp. (in Russian).

Kuznaik E., Sklodowska M. (2005). Fungal pathogen-induced changes in the antioxidant systems of leaf peroxisomes from infected tomato plants. *Planta*, v.222, p.192-200.

Laemmli U. (1970). Cleavage of structural proteins during the assembly of the head of bacteriophage T_4. *Nature*, v.227, p.680-685.

Lafitte, H.R., Yongsheng G., Yan S. and Li Z.K., (2007). Whole plant responses, key processes and adaptation to drought stress: The case of rice. *J. Exp. Bot.*, v.58, p.169-175.

Maksimov I.V., Cherepanova E.A. (2006). Pro/antioxidant system and plant tolerance to pathogens. *Achievements of the modern biology*, v.126, p.250-261 (in Russian).

Mayer A.M., Staples R.C., Gil-ad N.L. (2001). Mechanisms of survival of necrotrophic fungal plant pathogens in hosts expressing the hypersensitive response. *Phytochemistry*, v.5, p.33-41.

Merzlyak M.N. (1989). Activated oxygen and oxidative processes in plant cell membranes. *Results of the science and technique* (ser. Plant physiol). M.: RISTI, v.6, 167 pp. (in Russian).

Minibayeva F.V., Gordon L.Ch. (2003). Superoxide production and activity of extracellular peroxidase in plant tissues under stress. *Plant physiology*, v.50, p.459-464 (in Russian).

Mitteler R. (2002). Oxidative stress, antioxidants and stress tolerance. *Trends in Plant Science,* v.7 (9), p.405-410.

Mittler R., Zilinskas B. (1992). Molecular cloning and characterization of a gene encoding pea cytosolic ascorbate peroxidase. *J. Biol. Chem.*, v.267, p.21802-21807.

Mittler R., Zilinskas B. (1993). Detection of ascorbate peroxidase activity on native gels by inhibition of the ascorbate dependent reduction of nitro blue tetrazolium. *Anal. Biochem.*, v.212, p.540-546.

Mittova V., Tal M., Volokita M., Guy M. (2003). Up-regulation of the leaf mitochondrial and peroxisomal antioxidative systems in response to salt-induced oxidative stress in the wild salt-tolerant tomato species *Lycopersicon pennellii*. *Plant Cell Envir.*, v.26, p.845-856.

Mohamed A.M., Raldugina G.H., Kholodova V.P., Kuznetsov V.V. (2006). Accumulation of osmolytes in different genotypes of canola plants under chloride salinity. *Plant Physiology*, v.53, p.732-739 (in Russian).

Mubarakshina M.M., Khorobrykh S.A., Kozuleva M.A., Ivanov B.N. (2006). *Intermembrane formation of hydrogen peroxide at oxygen reduction in thylakoids of higher plants.* Reports of RAS, v.408(1), p.118-121 (in Russian).

Nagy N.S., Fossdal.G., Dalen L.S., et al. (2004). Effect of rhizoctonia infection and drought on peroxidase and chitnase activity in Norway spruce (*Picea abies*). *Physiol. Plant.*, v.120, p.465-473.

Navari-Izzo F., Quartassi M., Pinzino C. et al. (1998). Thylakoid-bound and stromal enzymes in wheat treated with excess copper. *Physiol. Plant.*, 104, 630-638.

Oberschall A., Deak M., Torok K., et al. (2000). A novel aldos/aldehyde reductase protects transgenic plants against lipid peroxidation under chemical and drought stress. *Plant J.*, v.24, p.437-446.

Park H.M., Kim Y.H., Shin D.B. et al. (2008). Enhanced tolerance to various abiotic stresses in transgenic rice by expressing *PsAPX* gene. *Physiol. Plant.*, 133, Abstract of the XVI Congr. of the FESPB, P07-050.

Passioura J.B., Angus J.F. (2010). Improving productivity of crops in water-limited environments. *Advances in Agronomy*, v.106, p.37-75.

Polesskaya O.G. (2007). *Plant cell and active oxygen forms.* M.: KDU, 140 pp. (in Russian).

Rogozhin V.V. (2004). *Peroxidase as a component of living organisms.* V.V. Rogozhin-SPb.: GIORD, 240pp. (in Russian).

Rogozhin V.V. (2010*). Plant peroxidase*. Berlin: Lambert Academic Publishing, 205 pp. (in Russian).

Sayar R., Khemira H., Kameli A., Mosbahi M. (2008). Physiological tests as predictive appreciation for drought tolerance in durum wheat (*Triticum durum* Desf.). *Agronomy research*, v.6 (1), p.79-80.

Scandalios J.G. (1993). Oxygen stress and superoxide dismutase. *Plant Physiol.*, v.101, p.7-12.

Sedmak J.J., Grossberg S.E. (1977). A rapid, sensitive and versatile assay for protein using Coomassie brilliant blue G 250. *Anal. Biochem.*, v.79, p.544-552.

Selate D.S., Khanna-Chopra R. (2006). Drought acclimation confers oxidative stress tolerance by inducing coordinated antioxidant defense at cellular and subcellular level in leaves of wheat seedlings. *Physiol. Plant.*, v.127, p.494-506.

Shi W.M., Muramoto Y., Ueda A. and Takabe T. (2001). Cloning of peroxisomal ascorbate peroxidase gene from barley and enhanced thermotolerance by overexpressing in *Arabidopsis thaliana*. *Gene*, v.273, p.23-27.

Shigeoka S., Ishikawa T., Tamoi M., Miyagawa Y., Takeda T., Yabuta Y. and Toshimura K. (1980). Regulation and function of ascorbate peroxidase isoenzymes. *J. Exp. Bot.*, v.53, p.1305-1319.

Strange R.N. and Scott P.R. (2005). Plant disease: A threat to global food security. *Annu. Rev. Phytopathol.*, v.43, p.83-116.

Takase M., Sam-Amoah L.K. and. Owusu-Sekyere J.D. (2011). The effects of four sources of irrigation water on soil chemical and physical properties. *Asian J. Plant Sci.*, v.10, p.92-96.

Takeda T., Yoshimura K., Yoshii M., Kanahoshi H., Miyasaka H. and Shigeoka S. (2000). Molecular characterization and physiological role of ascorbate peroxidase from halotolerant *Chlamydomonas* sp. W80 strain. *Arch. Biochem. Biophys.*, v.376, p.82-90.

Tang W., Charles T.M., Newton R.J. (2005). Overexpression of the pepper transcription factor CaPF1 in transgenic Virginia pine (*Pinus virginiana* Mill.) confers multiple stress tolerance and enhances organ growth. *Plant Mol. Biol.,* v.59 (4), p.603–617.

Tarchevskiy I.A. (2002). *Signaling systems in plant cells.* M. Nauka, 294 pp. (in Russian).

Tas S. and Tas B. (2007). Some physiological responses of drought stress in wheat genotypes with different ploidity in Turkey. *World Journal of Agricultural Sciences,* v.3, p.178-183.

Verma S. and Dubey R.S. (2003). Lead toxicity induces lipid peroxidation and alters the activities of antioxidant enzymes in growing rice plants. *Plant Sci.,* v.164, p.645-655.

Waterworth H.E. and Hadidi A. (1998). Economic Losses Due to Plant Viruses. In: *Plant Virus Disease Control*, Hadidi, A., R.K. Khetarpal and H. Koganezawa (Eds.). American Phytopathological Society Press, St. Paul, Minnesota, p.1-13.

Wilkinson S.R., Obado S.O., Mauricio I.I., Kelly J.M. (2002). *Trypanosoma cruzi* expresses a plan-like ascorbate-dependent hemoperoxidase localized to the endoplasmic reticulum. *Proc. Natl. Acad. Sci.* USA, v.99, p.13453-13458.

Yabuta Y., Motoki T., Yoshimura K., Takeda T., Ishikawa T. and Shigeoka S. (2002). Thylakoid membrane-bound ascorbate peroxidase is limiting factor of antioxidative systems under photo-oxidative stress. *Plant J.,* v.32, p.915-925.

Zhang J.X. and Kirkham M.B. (1996). Enzymatic responses of the ascorbate-glutathione cycle to drought in sorghum and sunflower plants. *Plant Sci.* v.113, p.139-147.

Zhao J., Sakai K. (2003). Multiple signaling pathways mediate fungal elicitor-induced β-thujaplicin biosynthesis in *Cupressus lusitanica* cell cultures. *J Exp Bot.*, v.54, p.647–656.

In: Peroxidases
Editors: Lucas Bogaert and Noah Coppens
ISBN: 978-1-62808-261-6

Chapter VII

Route, Dose and Duration of Exposure to Cadmium-Relevance to Oxidative Stress Induction

Vesna Matović*, Danijela Đukić-Ćosić, Aleksandra Buha and Zorica Bulat
Department of Toxicology "Akademik Danilo Soldatović",
University of Belgrade-Faculty of Pharmacy, Belgrade, Serbia

Abstract

Cadmium (Cd) is a toxic metal currently ranked as seventh (out of 275) on ATSDR priority list of hazardous substances and it is considered as one of the most important occupational and environmental pollutants. This carcinogenic metal causes adverse effects in various tissues, particularly kidneys after prolonged exposure. However, the mechanisms of these toxic effects are still not completely understood. Literature data indicate different mechanisms of cadmium toxicity, oxidative stress being one of the most important. Experiments confirm that Cd, metal with no redox potential, can indirectly induce the production of reactive oxygen species (ROS) by affecting antioxidant enzymes (superoxide dismutase, catalase, glutathione peroxidase, glutathione reductase, glutathione S-transferase), nonenzymatic antioxidants (glutathione, total sulfhydryl groups, antioxidative vitamines), and Fenton metals, iron and copper.

This chapter summarizes our investigations on cadmium-induced oxidative stress conducted on three different animal species: rabbits, mice and rats. Application of a single oral dose of 50 mg Cd/kg b.w. in rabbits showed increased levels of lipid peroxidation in blood 4-6 hours after Cd intoxication indicating early development of oxidative stress in conditions of acute Cd poisoning. Experiments carried out on mice showed that single oral exposure to 20 mg Cd/kg b.w. resulted in intensive ROS production in liver while subacute oral dose of 10 mg Cd/kg b.w. induced more pronounced effect in kidneys, thus indicating different effect of Cd intoxication

* Corresponding author Email: vevodi@pharmacy.bg.ac.rs.

depending on duration of Cd exposure. Further experiments performed on rats provided strong evidence that Cd can induce oxidative stress in both blood and liver after oral (30 mg Cd/kg b.w.) and intraperitoneal (1.5 mg Cd/kg b.w.) acute intoxication, although intraperitoneal administration showed more pronounced negative effects.

Results of these studies demonstrate that oxidative stress is an important mechanism of Cd toxicity and that Cd ability to induce oxidative stress in different tissues depends on the route, dose and duration of exposure to this toxic metal.

Cadmium-Ubiquitous Toxic Metal of Nowadays

Over the last few decades, cadmium (Cd) has been recognized as one of the most important environmental and industrial pollutants and predictions are that the exposure to this metal will be even increased in future decades (Nawrot et al., 2008).

During the mining of ores and ore extraction processes, this naturally occurring element can contaminate environment (Casado et al., 2008). Occupational exposures to cadmium also include nickel-cadmium batteries industry, electroplating and paint pigments. Other anthropogenic sources such as application of phosphate fertilizers, sewage sludge to land, the disposal of wastewaters, and smelter emissions significantly contribute to soil contamination by this metal (ATSRD, 2012), thus increasing Cd uptake into crops and green leafy vegetables.

Food is the major source of Cd exposure for general population with average Cd intake via food varying from 8 to 25 μg/day (Järup and Åkesson, 2009). It was estimated that more than 80% of the food-cadmium intake comes from cereals and vegetables, although some kinds of food such as sea food (molluscs and crustaceans), as well as offal products such as liver and kidney from older animals contain high levels of Cd. However, in Cd contaminated areas such as those in the vicinity of cadmium emitting plants or mining districts, ground water and surface water as well as atmosphere may be substantially contaminated by Cd.

Tobacco smoking is another source of cadmium exposure since tobacco leaves can accumulate this toxic metal. One cigarette contains about 1-2 μg Cd and in heavy smokers cadmium uptake from inhaling cigarette smoke can even exceed the uptake from all other sources of Cd (Järup and Åkesson, 2009).

Discovered in 1817, the first reports on adverse effects of Cd were given by Friberg in the middle of last century indicating that damage to lungs and kidneys might be the earliest effects and “critical effects” occurring in occupational Cd exposure.

A few years later itai-itai disease confirmed adverse effect of Cd on bones (Nordberg, 2009). As recently reviewed by Matović et al. (2011), nowadays it is known that Cd can also cause toxic effects on liver, pancreas, reproductive organs, placenta, and hematopoietic, nervous, and cardiovascular systems, with pulmonary system being the target of Cd associated carcinogenesis. Thus, the International Agency for Research on Cancer and the US National Toxicology Program classified Cd as human carcinogen (IARC, 1993; NTP, 2011). However, recent investigations and epidemiological studies confirmed that even low levels of Cd due to its very long biological half-life (even 30 years in kidneys) may be associated with pre-diabetes and diabetes mellitus, myocardial infarction, osteotoxicity, and distruption of different endocrine functions (Matović et al., 2011; ATSDR, 2012). Recent studies have also shown that Cd can alter thyroid function (Hammouda et al., 2008) and act as a metalloestrogen (Silva et al., 2012).

As a toxic metal with a wide range of organ toxicity, cadmium is expected to exhibit different mechanisms of toxicity. Majority of authors agree that the most important molecular mechanisms by which cadmium manifests its toxic effects and its carcinogenicity include changes in gene expression, inhibition of damaged DNA repair, inhibition of apoptosis, induction of oxidative stress, and interactions with bioelements (Joseph, 2009; Liu et al., 2009; Matović et al., 2010; Moulis, 2010).

Although over the last few decades substantial number of investigations has been conducted in order to precisely define mechanisms of Cd toxicity, further investigations are still necessary to provide additional information for understanding the complexity of Cd toxicity. However, it seems that induction of oxidative stress can have impact and important role in all other mechanisms of Cd toxicity.

Oxidative Stress in Cadmium-Induced Toxicity

Toxicity of metals, especially those with variable valence, is often associated with oxidative tissue damage. Although Cd is not a redox-active metal such as iron (Fe), copper (Cu) and chromium (Cr), it has been shown in *in vivo* and *in vitro* studies conducted in recent years that Cd can cause the production of reactive oxygen species (ROS) and reactive nitrogen species (RNS). Indirect formation of ROS and RNS involving the superoxide radical ($O_2^{\bullet -}$), hydroxyl radical ($OH^{\bullet}$) and nitric oxide (NO) has been reported by many authors (Buha et al., 2012; Djukić-Ćosić et al., 2008; Liu et al., 2009; Matović et al., 2012; Yang et al., 2007; Waisberg et al., 2003). The concentration of ROS and RNS can be increased either by their overproduction or by an inability of antioxidant system to destroy them. Numerous experiments were performed in order to clarify Cd effects on antioxidants enzymes (superoxide dismutase (SOD), catalase, glutathione peroxidase (GPX), glutathione reductase, glutathione S-transferase) and nonenzymatic antioxidants (glutathione (GSH), total sulfhydryl groups (TSH), antioxidative vitamines) in various organs. These studies, although not coherent, indicate cadmium ability to interfere with components of antioxidant system. Another mechanism involved in the indirect role of cadmium in free radical generation is that cadmium can replace Fe and Cu in various cytoplasmic and membrane proteins (e.g. ferritin, apoferritin), and thus increase the amount of these metals in unbound ionic state that can induce oxidative stress via Fenton reactions (Casalino et al., 1997).

Since the role of oxidative stress in Cd toxicity still appears rather controversial, our researches were focused on this particular mechanism of Cd toxicity. In this chapter we will summarize the results of our experiments performed on three different animal species: rabbits, mice and rats. During these investigations different doses of Cd were administrated by different routes of exposure while duration of exposure to this toxic metal was either acute or subacute. Different parameters of oxidative stress were determined during these researches in order to get better insight into Cd mechanisms of oxidative stress induction. Concentrations of following reactive species were determined: $O_2^{\bullet -}$ and NO and total oxidative status (TOS) which besides $O_2^{\bullet -}$ levels include other reactive oxygen metabolites such as H_2O_2 and lipid hydroperoxides (LOOH). Furthermore, malondialdehyde (MDA) content as a measure of lipid peroxidation and advanced oxidation protein products (AOPP) were determined. Among the parameters of antioxidative defense system, the activity of antioxidant enzyme SOD, as well as the content of nonenzymatic antioxidants TSH and GSH were determined.

Cadmium-Induced Oxidative Stress in Rabbits-Our Pioneer Investigation

Our investigations on Cd-induced oxidative stress started in 90s and were performed on rabbits *Oryctolagus cunniculus-Belgian hare* given single oral dose of 50 mg Cd/kg b.w. as aqueous solution of $CdCl_2$. Blood samples were taken from ear artery before intoxication and 1, 3, 4, 6, and 24 hours after the treatment. Levels of $O_2^{\bullet-}$ were determined as a rate of nitrobluetetrazolium (NBT) reduction according to Auclair and Voisin (1985), free MDA level was determined by the method proposed by Girotti et al. (1991), while stimulated MDA level was measured after stimulation with $FeSO_4$ by method given by Andreeva (1988). All parameters were determined in plasma. The results showed the increased generation of $O_2^{\bullet-}$ as a parameter of the earliest phase of oxidative stress. Significant elevation of $O_2^{\bullet-}$ was observed 4 hours after Cd administration, if compared with controls not treated with Cd. The increased level in both free and stimulated MDA developed also 4-6 hours after acute intoxication with Cd demonstrating an intensive lipid peroxidation and development of oxidative stress shortly after single treatment with Cd (Matović et al., 1998; Plamenac et al., 2002).

Cadmium-Induced Oxidative Stress in Mice-Influence of Exposure Duration

The aim of our further investigation was to evaluate the influence of duration of Cd exposure on parameters of oxidative stress in mice liver and kidneys as two target organs of Cd toxicity. The first part of this research was a time-course study carried out on 40 *Swiss albino* mice given a single oral dose of 20 mg Cd/kg b.w. by orogastric tube as aqueous solution of $CdCl_2$. The other one was performed on 16 *Swiss albino* mice orally treated with 10 mg Cd/kg b.w. every day for 14 days. In both acute and subacute intoxication MDA content, NO production, total SOD and Cu, Zn-SOD activity, as well as GSH and TSH content as parameters of oxidative stress were determined in liver and kidneys. Malondyaldehide content was determined by method of Girotti et al. (1991), NO production by modified Griess method given by Cortasa and Waklda (1990), SOD and Cu, Zn-SOD activity was determined by method proposed by Misra and Fridovich (1972) and GSH and TSH contents were determined by Ellman method (1959). These parameters were measured 4, 6, 12, 24, and 48h after single Cd treatment and on 7th and 14th day of subacute Cd intoxication (groups of 8 animals each). One group of eight animals was used as the control (untreated animals) and was sacrificed at time zero. Statistical analysis was performed by ANOVA followed by LSD post-hoc test and $P<0.05$ were considered as statistically significant.

Our results suggest development of early oxidative stress in liver of mice after acute intoxication with Cd, since significant increase of hepatic MDA was observed after 6, 12 and 24h (Djukić-Ćosić et al., 2008). These findings are in agreement with investigations performed by Eybl et al. (2004) who also found elevated MDA levels in liver of mice 24 hours after treatment with single subcutaneous dose of 33 μmol $CdCl_2$/kg b.w. The observed Cd-induced lipid peroxidation in liver is rather difficult to explain since Cd has no redox activity. However, since previous studies (Casalino et al., 1997) showed that Cd can cause redistribution of iron (Fe) and replacement of Fe in Fe-dependant enzymes and proteins, we extended our investigations on the effect of Cd on hepatic levels of this Fenton metal. Under the same experimental conditions statistically significant positive correlation between degree of Fe concentration and MDA in mouse liver was found across the time points for Fe (0, 4, 6, 12h) and for MDA (4, 6, 12, 24h) with $\rho=0.44$, $P=0.026$ (Djukić-Ćosić et al., 2008). These data confirm that Cd can indirectly induce production of ROS by its interaction with Fe, namely increasing the concentration of ionic Fe which as a Fenton metal stimulates free radicals production and oxidative damage of lipids. However, it should be pointed that acute Cd intoxication even caused reduction of hepatic MDA if compared with control levels at the end of experiment (48h). This finding could be explained by the development of delayed adaptive response of the organism to Cd exposure. When concerning kidneys, the effect of acute Cd intoxication on MDA levels was less pronounced and its statistically significant increase was observed only 12h after Cd treatment.

Production of RNS under the same experimental conditions was evaluated by determination of nitrate content in liver and kidneys. The only significance was observed in liver after 12h when nitrate levels were elevated when compared to control levels (Đukić-Ćosić et al., 2010; Đukić-Ćosić, 2011).

Numerous literature data indicate that Cd induces oxidative stress indirectly by affecting antioxidant defense system. Our results have shown that acute Cd intoxication induces a significant decrease of hepatic and renal total and Cu,Zn-SOD activity after 6h which lasted until the end of experiment if compared with controls. Decrease of Cu,Zn-SOD activity in mice kidneys was accompanied by reduced Cu and zinc (Zn) content in mice kidneys, but no significant alteration in Zn levels was observed in liver (Đukić-Ćosić et al., 2009; Đukić-Ćosić et al., 2012). These findings indicate possible replacement of Cu and Zn by Cd in kidneys, while reduced activity of this enzyme in liver could be the consequence of direct effect of high levels of Cd determined in this organ (Djukić-Ćosić et al., 2006; Đukić-Ćosić et al., 2012).

No changes in total SH levels were observed either in liver or in kidneys of mice exposed to acute Cd intoxication. However, single exposure to Cd induced significant alterations of GSH content in both investigated organs, with reduction of GSH observed in liver and its increase observed in kidneys. Hepatic GSH content was diminished if compared with controls after 4, 6 and 12h while statistically significant elevation of GSH level was obtained in kidneys after 12, 24 and 48h. In general, acute exposure to metals results in decrease of GSH levels as a consequence of metal-GSH complexes formation and/or GSH consumption by GPX which is elevated in oxidative stress. Similar observations were given by some other authors (El-Maraghy et al., 2001; Eybl et al., 2004). Predominant accumulation of Cd in liver if compared to kidneys could be also explanation for favorable inhibitory effect of Cd on hepatic GSH content during the first 12h of our experiment. Normal levels of hepatic GSH levels reached after 24 and 48h could be explained by *de novo* GSH synthesis and by enhanced Cd elimination (Đukić-Ćosić, 2011). However, it is rather difficult to explain the effect of Cd on renal GSH content and it remains to be answered whether applied dose of Cd induces raised levels of GSH as a delayed activation of organism defense mechanisms (Djukić-Ćosić et al., 2007). The most pronounced effects of Cd acute intoxication on parameters of oxidative stress in mice liver are presented in Figure 1.

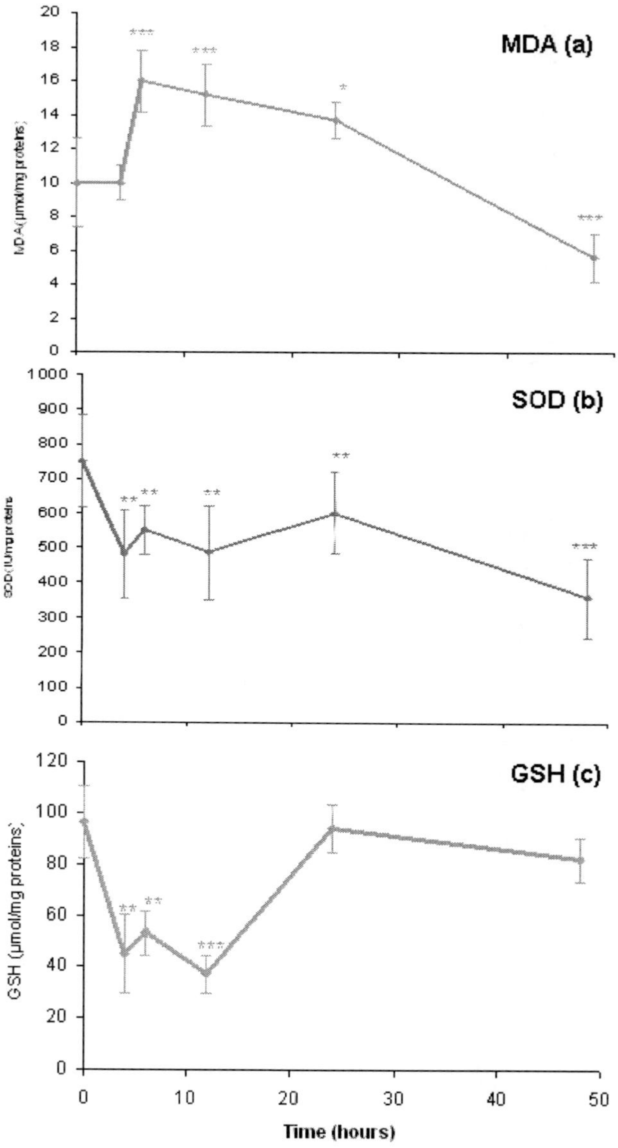

Figure 1. The effects of single oral dose of 20 mg Cd/kg b.w. on a) MDA content, b) SOD activity and c) GSH content in mice liver.

Values are presented as the means ± SD. Marked values differ significantly (ANOVA/LSD) from control group (zero time). **P*<0.05; ***P*<0.01; ****P*<0.001.

Results obtained for subacute Cd intoxication indicate less pronounced effect on parameters of oxidative stress in liver if compared to the results obtained for acute poisoning.

Levels of MDA and nitrates were even decreased after one and two weeks treatment with Cd. The observed decrease in MDA might be the result of different mechanisms. Boujelben et al. (2006) demonstrated that decreased lipid peroxidation in the liver after intoxication with 2.5 mg $CdCl_2$/day/kg b.w. for 10 days is a result of the activation of antioxidant defense mechanisms.

This mechanism was also proposed by Klaassen et al. (1999, 2009), Nordberg (1999) and Nordberg and Nordberg (2000) who showed that Cd induced synthesis of hepatic metallothioneins as efficient scavengers of Cd and free radicals.

Increased levels of GSH accompanied with elevated TSH content in hepathocytes observed in our study after one and two weeks (statistically significant after two weeks), which was also observed by El-Maraghy et al. (2001), could be the other possible explanation for protective mechanism against Cd-induced oxidative stress.

However, our opinion is that this phenomenon could be mainly attributed to depleted Fe content in liver during subacute exposure to Cd, since the correlation between MDA and Fe concentration was positive and statistically significant during the whole period of subacute experiment (ρ=0.69, P=0.003) (Djukić-Ćosić et al., 2008).

On the other hand, investigated Cu, Zn-SOD and total SOD in liver were in the levels of controls during the entire experiment. However, two weeks treatment with 10 mg Cd/kg b.w. resulted in oxidative and nitrosative stress in kidney of mice. The enhanced free radicals formation in kidneys was accompanied and can be explained by elevated levels of Fenton metals Fe and Cu, increased activity of SOD and diminished GSH and TSH levels (Đukić-Ćosić, 2011; Đukić-Ćosić et al., 2011). As suggested by Karmakar et al. (1998), the decrease in the GSH level might be a consequence of the increased activity of γ-glutamyl transpeptidase enzyme, an external cell-surface enzyme that can enhance the degradation of extracellular GSH.

Another possible reason for decreased GSH level may be attributed to increased levels of lipid oxidation products which could be associated with the less availability of NADPH required for glutathione reductase activity to transform oxidized glutathione (GSSG) to the reduced form (GSH) (Sarkar et al., 1995). The effects of subacute Cd intoxication on the levels of parameters of oxidative stress in kidneys of mice are summarized in Table 1.

Based on the presented results, it can be concluded that Cd treatment stimulates the production of the major mediators of oxidative damage, ROS and RNS, both in liver and kidneys. However, Cd behavior concerning its effect on oxidative stress was different in the conditions of acute and subacute poisoning. Mice exposed to single oral Cd treatment exerted more marked effects in liver than in kidneys reflected by the elevation of MDA levels and decreased activity of investigated antioxidant enzymes and GSH as a nonenzymatic component of antioxidative defense.

These effects were observed very shortly after Cd treatment (6h) suggesting that Cd exposure rapidly disrupts oxidative-antioxidative balance in liver through already suggested mechanisms. It should be emphasized once again that enhanced production of free radicals could be explained by Cd interactions with endogenous Fe since significant Fe elevation in liver was induced by Cd.

Table 1. The effect of two weeks Cd treatment (10 mg Cd/kg b.w.) on the MDA content, SOD activity, GSH and TSH content in mice kidneys

	Time (days)		
	control	7 days	14 days
MDA *(μmol/mg proteins)*	5.96 ± 0.48	5.91 ± 0.94	8.07 ± 1.55**
SOD *(IU/mg proteins)*	294.81 ± 38.37	258.81 ± 52.80	401.01 ± 76.36**
GSH *(μmol/mg proteins)*	11.62 ± 2.46	7.63 ± 2.15**	9.48 ± 2.98**
TSH *(mmol/mg proteins)*	80.65 ± 11.10	88.10 ± 19.39	67.10 ± 2.50

Values are expressed as mean±SD. Asterisk (*) denotes significant difference from the control group ($P<0.05$); double asterisk (**) denote significant difference from the control group ($P<0.01$) (one-way ANOVA plus LSD post-hoc test).

Conversely, following subacute Cd intoxication increased ROS and RNS levels were observed in kidneys, but not in liver. Indeed, kidneys are generally regarded as the critical organ of long-term exposure to Cd. The induction of metallothioneins which can sequester Cd is one of the plausible explanations for the amelioration of investigated parameters levels in liver after subacute intoxication.

The results of this investigation suggest that cadmium has significant effect on oxidative stress by generating ROS and RNS through depletion of endogenous radical scavengers and interactions with Fe, which in turn leads to various cell damages and significantly contributes to hepatotoxicity and nephrotoxicity of Cd. In addition, our results highlight the different effects of acute and subacute Cd intoxication on investigated parameters of oxidative stress in liver and kidneys suggesting dependency of these effects on duration of exposure to this toxic metal.

Cadmium-Induced Oxidative Stress in Rats-Influence of Route of Exposure

Our most recent study was aimed to determine how different routes of acute exposure to Cd affect parameters of oxidative stress in blood and liver of rats. Investigated routes of exposure were oral (*or*) and intraperitoneal *(i.p.)* exposure. These two particular routes of exposure were chosen based on two possible scenarios of human exposure to Cd: a) *or*-corresponding to exposure through food and water and b) *i.p.*- corresponding to parenteral exposure to Cd as a result of occupational Cd exposure via lungs and/or inhalation of tobacco smoke in general population.

The animals used in this study were randomized into three groups of eight rats each. For *or* treatment, aqueous solution of $CdCl_2$ was administrated in a single dose of 30 mg Cd/kg b.w. by orogastric tube. Intraperitoneally treated rats were injected an isotonic $CdCl_2$ solution in a single dose of 1.5 mg Cd/kg b.w. Control animals were unexposed through the whole experiment. Twenty four hours after intoxication animals were sacrificed under ether anesthesia; blood samples were collected from heart, and livers were excised. After separating plasma from blood and liver homogenization, following parameters of oxidative stress were analyzed: $O_2^{\bullet-}$ and MDA concentrations, SOD activity, AOPP, TOS, TSH and GSH content. The rate of $O_2^{\bullet-}$ formation, MDA concentration, SOD activity in blood and liver and GSH and TSH groups content in liver homogenates were determined by the same methods used in our investigations on rabbits and rats, described above. Levels of AOPP as indicators of ROS-mediated protein damage were measured in plasma by method of Witko-Sarsat et al. (1991). Total oxidative status ($O_2^{\bullet-}$, H_2O_2 and LOOH levels) was assayed in plasma using the method described by Erel (2005). The results obtained in plasma were statistically analyzed by Kruskal-Wallis nonparametric test followed by post-hoc Conover test for pairwise comparison, while results obtained in liver were analyzed using ANOVA followed by post-hoc LSD test and $P<0.05$ were considered as statistically significant.

The effects of Cd treatment on investigated parameters of oxidative stress in plasma are represented in Table 2. Cadmium *or* and *i.p.* administration induced a significant increase in both $O_2^{\bullet-}$ and TOS plasma levels when compared to controls, suggesting Cd ability to induce increase in plasma levels of ROS. Investigations of other authors also indicated that exposure to Cd can elicit significant $O_2^{\bullet-}$ generation, as in a study conducted by Filho et al. (2000) who treated rats with a single intraperitoneal dose of 2 mg Cd/kg. In our investigation conducted on rabbits this increase was observed even 4h after Cd treatment, suggesting that this Cd effect can be regarded as an early marker of Cd induction of oxidative stress. Cadmium effect on $O_2^{\bullet-}$ generation can be explained by Cd adverse effect on SOD activity reflected by its decrease.

Table 2. The effect of Cd treatment on the SOD activity, $O_2^{\bullet-}$, TOS and MDA levels in plasma

Parameter		Control	Cd or	Cd ip
$O_2^{\bullet-}$ *(μmol red. NBT/min/L)*	Median	87.5	107^{*}	$284^{*,\dagger}$
	Range	60-99	101-128	234-370
TOS *(μmol/L)*	Median	28.3	49^{*}	$68.3^{*,\dagger}$
	Range	15.6-39.8	41.6-63.2	46-83.8
MDA *(μmol/L)*	Median	0.68	1.14^{*}	$1.82^{*,\dagger}$
	Range	0.46-0.93	0.93-1.69	1.36-1.93
SOD *(IU)*	Median	137	130^{*}	123^{*}
	Range	131-139	127-134	119-133

Control group-non-treated animals, Cd_{or} group- rats intoxicated orally with 30 mg Cd/kg b.w.; $Cd_{i.p.}$ group-rats intraperitoneally injected a single dose of 1.5 mg Cd/kg b.w. The values are expressed as medians and ranges. Statistically significant differences (Kruskal-Wallis nonparametric test, Conover test for pairwise comparison) obtained for each parameter between groups are indicated by: *vs. control, †vs. Cd_{or}. $P<0.05$.

To our knowledge, levels of TOS with a reference to Cd intoxication have not been studied up to date. Obtained increase of TOS levels can also be the illustration of Cd ability to decrease the activity of enzymatic components of antioxidative protection. Gluthatione peroxidase plays an important role in LOOH reduction to corresponding alcohols and in the degradation of H_2O_2 to H_2O. Hence, increased levels of TOS observed in this study can be explained by adverse effect of Cd on GPX activity in plasma of treated rats, since some other authors announced decrease of GPX activity in liver in conditions of acute intraperitoneal intoxication (Liu et al., 2011) and subchronic Cd oral intoxication in serum, liver and kidney (Galazyn-Sidorczuk et al., 2012). Furthermore, significant increase in plasma levels of both $O_2^{\bullet-}$ and TOS in our study was much more marked in the group that received Cd intraperitoneally than in a group treated orally.

During oxidative stress, besides eliciting widespread damage to cells such as lipid peroxidation of polyunsaturated membrane lipids leading to MDA generation, ROS can also produce oxidative damage of proteins that can be marked by AOPP levels. Both *or* and *i.p.* Cd treatment resulted in a significant increase in MDA and AOPP when compared to controls, and these levels were significantly higher in the group treated intraperitoneally than in orally treated one. Similar results were acknowledged by some other authors. Thus, subcutaneous daily treatment of rats with 0.49 mg Cd/kg for 20 days and intraperitoneal daily treatment with 2 mg $CdCl_2$/kg during 10 days significantly increased MDA plasma levels (Nemmiche et al., 2007; Kanter et al., 2009). Although no literature data on the effect of Cd on AOPP levels are available, Gałazyn-Sidorczuk et al. (2009) investigated generated damage of proteins by investigating the concentration of protein-bound carbonyl groups in serum. In this experiment conducted on rats administration of Cd by drinking water containing 50 mg Cd/L induced significant increase of carbonyl serum levels after 12 weeks. In our acute studies on rabbits and rats, elevated plasma MDA and AOPP levels were obtained much earlier i.e. within 24 hours, probably as a result of different applied Cd doses.

As it was expected, Cd treatment decreased SOD activity in plasma if compared to the controls, but surprisingly no significant difference between SOD activity after *or* and *i.p.* administration of Cd was observed. Cadmium-incited decrease in SOD activity could be explained by previously confirmed antagonism between Cd and Zn (Bulat et al., 2008; Bulat et al., 2012; Jihen et al., 2011). Thus, Bauer et al. (1980) reported that Cd was able to occupy the site of Zn in the Cu, Zn-SOD isoenzyme, creating an inactive form of this enzyme. Furthermore, our recent investigation (Buha et al., 2012) proved positive correlation between blood Zn and SOD activity (ρ=0.339, P=0.035) which was then confirmed for both oral and intraperitoneal Cd treatment (ρ=0.894, P=0.041 and ρ=0.721, P=0.042, respectively). However, intraperitoneal Cd administration resulted in more pronounced negative effects on all other investigated parameters. This route-dependency observed for Cd effects on investigated parameters in plasma can be explained by significantly higher blood Cd levels obtained after *i.p.* Cd administration as shown in our recent study (Matović et al., 2012).

Parameters of oxidative stress in rats treated with Cd orally and intraperitoneally were investigated not only in plasma, but in liver homogenates as well since liver is regarded as one of the target organs of Cd acute toxicity. Obtained results are summarized in Figure 2. Similarly to our previously described study on mice, single oral Cd treatment produced early marked effects on oxidative-antioxidative balance in liver. Increase of $O_2^{\bullet-}$ and MDA levels was significant after both oral and intraperitoneal treatment which is in agreement with some other authors results. Significantly increased MDA content was found in liver of rats sacrificed 24h after intraperitoneal treatment with single injection of 2.5 mg $CdCl_2$/kg b.w. (Casalino et al., 2002) and in liver of rats 16 hours after intraperitoneal treatment with 20 μmol Cd/kg b.w. when compared to controls (Liu et al., 2011). Besides MDA plasma levels, Nemmiche et al. (2007) investigated MDA levels in liver as well and found increased levels in Cd exposed group if compared with controls. Furthermore, in our study *i.p.* intake showed more articulate effect on these two parameters: levels of $O_2^{\bullet-}$ were almost 25% higher after *or* exposure and more than 50% higher after *i.p.* exposure when compared to controls, while MDA levels were 2.5 times higher in oral group and almost 3.5 times higher in intraperitoneal group than in control group. This route dependent effect on levels of both parameters can be contributed to significantly higher Cd content determined in liver after *i.p.* administration than after *or* administration (Matović et al., 2012). Indeed, correlation analysis showed strong positive correlations between Cd liver content and $O_2^{\bullet-}$ and MDA levels ($\rho = 0.703$, $P < 0.01$ and $\rho = 0.765$, $P < 0.01$, respectively).

Similarly to the results obtained in plasma of rats, SOD activity in liver was diminished with no statistically significant differences between SOD activities within groups treated orally or intraperitoneally. Decrease in liver activity of SOD shortly after Cd treatment was also observed in our study on mice, as well as in studies of other authors conducted on rats (Casalino et al., 2002; Yalin et al., 2005). As it has been previously proposed, interactions between Cd and bioelements Zn, Cu and Mn that are present in the SOD active center can be one of the explanations for the effect of Cd on SOD activity.

Besides enzymatic defense, we investigated TSH and GSH content in rats liver, as parameters of nonenzymatic defense system. Although considered to be important detoxifying factor against metal toxicity, TSH levels were significantly altered after Cd treatment, regardless the route of exposure. However, significant decrease of GSH, as one of the most important molecules containing SH groups, was observed after Cd *i.p.* treatment.

The explanation for different effect of *or* and *i.p.* Cd intoxication on GSH molecules content might be the assumption that GSH *de novo* synthesis could not overlap more pronounced reduction in GSH content after *i.p.* Cd administration when higher Cd levels in liver were determined (Matović et al., 2012). This is in accordance with our results obtained on mice and results of other authors (Đukić-Ćosić et al., 2007; El-Maraghy et al., 2001; Eybl et al., 2004).

The results of this study show significant decrease in SOD activity, which is not route dependent, but is accompanied with concurrent decline in GSH content after *i.p.* administration. This phenomenon can provide possible explanation for more profound lipid peroxidation observed in liver of intaperitoneally treated rats.

According to the results of our study performed on rats, it can be concluded that Cd induces oxidative stress in a route dependent manner. Elevated ROS levels were observed in both plasma and liver of orally and intraperitoneally treated rats. Further consequences related to ROS elevation, such as increased MDA and AOPP levels were also detected. Intraperitoneal which corresponds to parenteral exposure to Cd resulted in more serious damage of lipids and proteins than *or* exposure. This phenomenon can be attributed to higher Cd body burden observed after *i.p.* intoxication. Since Cd is not a redox-active metal, Cd-induced generation of ROS happens through depletion of enzymatic and non-enzymatic components of antioxidant defense system. Decrease in investigated enzymatic component of antioxidant defense (SOD activity) can be mainly attributed to displacement of metal co-factors from active site of SOD and surprisingly was not route-dependent.

On the other hand, decrease of GSH content as one of the important non-enzymatic component of antioxidant defense system was much more expressed after *i.p.* exposure. These observations suggest that non-enzymatic antioxidant system is more sensitive and more affected by higher Cd concentrations in the organism than enzymatic and/or can be ameliorated less effectively.

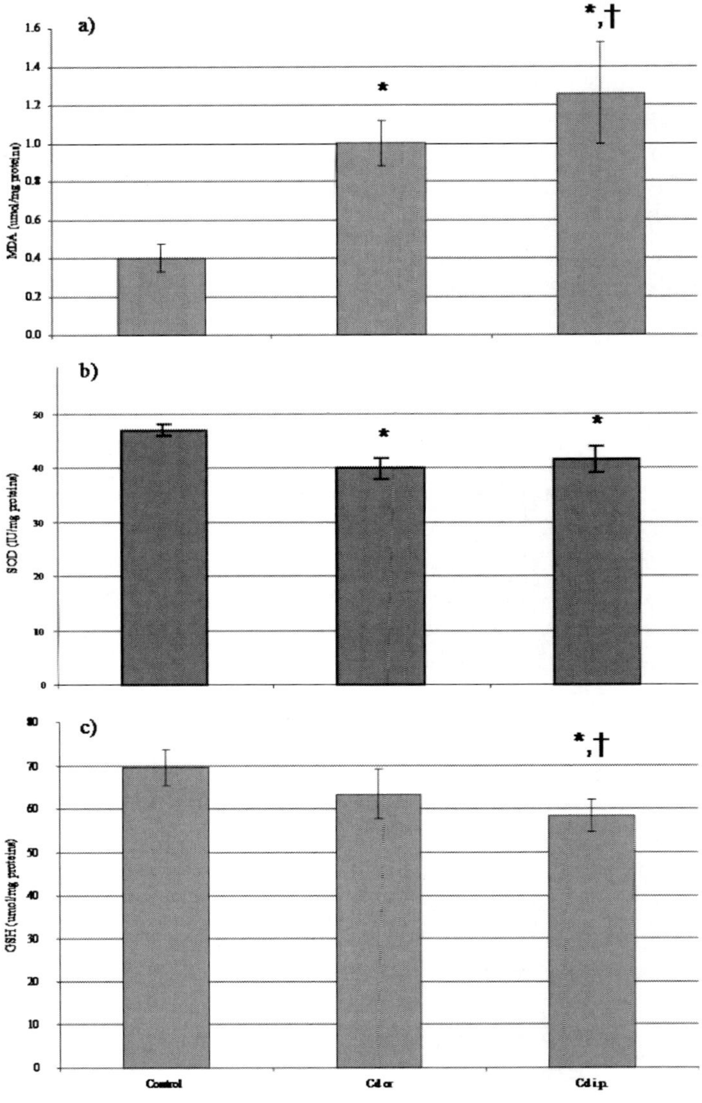

Figure 2. The effects of Cd treatment on a) MDA content, b) SOD activity and c) GSH content in liver of rats.

Control group-non-treated animals, Cd_{or} group-rats intoxicated orally with 30 mg Cd/kg b.w.; $Cd_{i.p.}$ group-rats intraperitoneally injected a single dose of 1.5 mg Cd/kg b.w. The values represent mean ± SD; n=8 animals. Statistically significant differences (ANOVA, LSD test) obtained for each parameter between groups are indicated by: *vs. control, †vs. Cd_{or}. $P<0.001$.

Conclusion

Our studies conducted on three different animal species with different dose levels of Cd provide substantial body of evidence that confirms oxidative stress as one of the mechanisms of Cd toxicity which can lead to serious and profound cell damages. Based on these results, lipid peroxidation can be regarded as early and sensitive consequence of acute Cd exposure since elevated MDA levels were obtained in all species within 24 hours of Cd administration. Duration of exposure was proved to be the major determinant for the critical organ of Cd toxicity, suggesting liver as an organ in which Cd rapidly exerts adverse effects after acute intoxication and kidneys as the critical organ of prolonged exposure to Cd. The route of exposure to Cd was also shown to play a crucial role in the oxidative potency of this toxic metal, demonstrating *i.p.* Cd administration as a route of exposure that has more pronounced adverse effects on majority of investigated parameters of oxidative stress in rat plasma and liver than the *or* one.

Acknowledgments

This work was partly supported by the Ministry of Education, Science and Technological Development of Serbia (Project III 46009).

References

Agency for Toxic Substances and Disease Registry (ATSDR), 2012. Toxicological profile for Cadmium. Atlanta, GA: U.S. Department of Health and Human Services, Public Health Service.

Auclair, C., Voisin, E., 1985. Nitroblue tetrazolium reduction. In: Greenwald RA, editor. Handbook of methods for oxygen radical research. *Boca Raton: CRC Press*; p. 123-32.

Andreeva, L. I., Kozhemiakin, L. A., Kishkun, A. A., 1988. Modification of the method of determining lipid peroxidation in a test using thiobarbituric acid. [Modifikatsiia metoda opredeleniia perekiseĭ lipidov v teste s tiobarbiturovoĭ kislotoĭ.] Laboratornoe Delo 11, 41-3.

Bauer, R., Demeter, I., Hasemann, V., Johansen, J. T., 1980. Structural properties of the zinc site in Cu,Zn-superoxide dismutase; perturbed angular correlation of gamma ray spectroscopy on the cu, 111Cd-superoxide dismutase derivative. *Biochem. Biophys. Res. Commun.* 94, 1296-302.

Beyersmann, D., Hartwig, A., 2008. Carcinogenic metal compounds: recent insight into molecular and cellular mechanisms. *Arch. Toxicol.* 82, 493-512.

Boujelben, M., Ghorbel, F., Vincent, C., Makni-Ayadi, F., Guermazi, F., Croute, F., El-Feki, A., 2006. Lipid peroxidation and HSP72/73 expression in rat following cadmium chloride administration: Interactions of magnesium supplementation. *Exp. Toxicol. Pathol.* 57, 437-43.

Buha, A., Bulat, Z., Đukić-Ćosić, D., Matović, V., 2012. Effects of Oral and Intraperitoneal Magnesium Treatment against Cadmium-Induced Oxidative Stress in Plasma of Rats. *Arh. Hig. Rada Toksikol.* 63, 247-53.

Bulat, Z., Đukić-Ćosić, D., Antonijević, B., Bulat, P., Vujanović, D., Buha, A., Matović, V., 2012. Effect of magnesium supplementation on the distribution patterns of zinc, copper, and magnesium in rabbits exposed to prolonged Cd intoxication. *ScientificWorldJournal* ID 572514, 9 pages, doi:10.1100/2012/572514.

Bulat, P.Z., Djukić-Ćosić, D., Maličević, Ž., Bulat, P., Matović V., 2008. Zn or Mg supplementation modulates Cd intoxication in blood, kidney, spleen and bone of rabbits. *Biol. Trace Elem. Res.* 124, 110-17.

Casalino, E., Calzaretti, G., Sblano, C., Landriscina, C., 2002. Molecular inhibitory machanisms of antioxidant enzymes in rat liver and kidney by cadmium. *Toxicology* 178, 37–50.

Casalino, E., Sblano, C., Landriscina, C., 1997. Enzyme activity alteration by cadmium administration to rats: the possibility of iron involvement in lipid peroxidation. *Arch. Biochem. Biophys.* 346, 171–9.

Casado, M., Anawar, H. M., Garcia-Sanchez, A., Santa Regina, I., 2008. Cadmium and zinc in polluted mining soils and uptake by plants. *Int. J. Environ. Poll.* 33, 146-59.

Cortas, N., Wakld, N., 1990. Determination of inorganic nitrate in serum and urine by akinetic cadmium/reduction method. *Clin. Chem.* 36, 1440-3.

Đukić-Ćosić, D., 2011. The effect of magnesium on oxidative stress and bioelements in mice exposed to acute and subacute cadmium intoxication. Doctoral disertation, University of Belgrade, Faculty of Pharmacy.

Djukić-Ćosić, D., Ćurčić Jovanović, M., Plamenac Bulat, Z., Ninković, M., Maličević, Ž., Matović, V., 2008. Relation between lipid peroxidation and iron concentration in mouse liver after acute and subacute cadmium intoxication. *J. Trace Elem. Med. Biol.* 22, 66-72.

Djukić-Ćosić, D,, Ninković, M., Maličević, Z., Matović, V., Soldatović, D., 2007. Effect of magnesium pretreatment on reduced glutathione levels in tissues of mice exposed to acute and subacute cadmium intoxication: a time course study. *Magnes. Res.* 20, 177-86.

Djukić-Ćosić, D., Ninković, M., Maličević, Z., Bulat-Plamenac, Z., Matović, V., 2006. Effect of supplemental magnesium in the kidney levels of cadmium, zinc, and copper of mice exposed to toxic levels of cadmium. *Biol. Trace. Elem. Res.* 114, 281-92.

Đukić-Ćosić, D., Bulat, Z., Buha, A., Ninković, M., Maličević, Ž., Matović, V., 2012. Mechanisms of Cu,Zn-SOD inhibition in tissues of mice exposed to acute Cd intoxication. Abstracts of the 48th Congress of the European Societies of Toxicology-EUROTOX 2012, Stockholm, Sweden, June 17-21 2012. *Toxicol. Lett.* 211, S142-S142.

Đukić-Ćosić, D., Vukomanović, P., Bulat, Z., Ninković, M., Maličević, Ž., Matović, V., 2011. Background explanation of oxidative stress in kidney of mice induced by subacute cadmium intoxication. Abstracts of the 47th Congress of the European Societies of Toxicology-EUROTOX 2011, Paris, France, August 2011. *Toxicol. Lett.* 205, S187-S187.

Đukić-Ćosić, D., Ninković, M., Maličević, Ž., Plamenac Bulat, Z., Matović, V., 2010. Effect of subacute cadmium intoxication on oxidative and nitrosative stress in mice. *Toxicol. Lett.* 196 (Supplement 1).

Đukić-Ćosić, D., Plamenac Bulat, Z., Ninković, M., Maličević, Ž., Matović, V., 2009. Renal superoxide dismutase activity in mice exposed to acute and subacute cadmium intoxication: The role of magnesium pretreatment. 46th Congress of the European Societies of Toxicology – EUROTOX 2009; September 13-16, 2009; Drezden, Germany; *Toxicol. Lett.* 189S, S222.

Ellman, G.L., 1959. Tissue sulfhidryl groups. *Arch. Biochem. Biophys.* 82, 70-7.

El-Maraghy, S.A., Gad, M.Z., Fahim, A.T., Hamdy, M.A., 2001. Effect of cadmium and aluminium intake on the antioxidant status and lipid peroxidation in rat tissues. *J. Biochem. Mol. Toxicol.* 15, 201-14.

Erel, O., 2005. A new automated colorimetric method for measuring total oxidant status. *Clin. Biochem.* 38,1103-11.

Eybl, V., Kotyzová, D., Bludovská, M., 2004. The effect of curcumin on cadmium-induced oxidative damage and trace elements level in the liver of rats and mice. *Toxicol. Lett.* 151, 79-85.

Filho Novelli, J. L. V. B., Novelli, E. L. B., Manzano, M. A., Lopes, A. M., Cataneo, A. C., Barbosa, L. L., Ribas, B. O., 2000. Effect of α-tocopherol on superoxide radical and toxicity of cadmium exposure. *Int. J. Environ. Heal. R.* 10, 125-34.

Gałazyn-Sidorczuk, M., Brzóska, M. M., Jurczuk, M., Moniuszko-Jakoniuk, J., 2009. Oxidative damage to proteins and DNA in rats exposed to cadmium and/or ethanol. *Chem. Biol. Interact.* 180, 31-8.

Galazyn-Sidorczuk, M., Brzóska, M. M., Rogalska, J., Roszczenko, A., Jurczuk, M., 2012. Effect of zinc supplementation on glutathione peroxidase activity and selenium concentration in the serum, liver and kidney of rats chronically exposed to cadmium. *J. Trace Elem. Med. Bio.* 26, 46-52.

Girotti, M., Khan, N., Mc Lellan, B., 1991. Early measurement of systemic lipid peroxidation products in the plasma of major blunt trauma patients. *J. Trauma* 31, 32–5.

Hammouda, F., Messaoudi, I., El Hani, J., Baati, T., Saïd, K., Kerkeni, A., 2008. Reversal of cadmium-induced thyroid dysfunction by selenium, zinc, or their combination in rat. *Biol. Trace Elem. Res.* 126, 194-203.

IARC, 1993. Cadmium and cadmium compounds, In: IARC Monographs on the evaluation of carcinogenic risks to humans. Vol. 58. Beryllium, cadmium, mercury, and exposures in the glass manufacturing industry. *IARC*, Lyon, pp.119–237.

Järup, L., Åkesson, A., 2009. Current status of cadmium as an environmental health problem. *Toxicol. Appl. Pharmacol.* 238, 201-8.

Jihen, E. H., Sonia, S., Fatima, H., Mohamed Tahar, S., Abdelhamid, K., 2011. Interrelationships between cadmium, zinc and antioxidants in the liver of the rat exposed orally to relatively high doses of cadmium and zinc. *Ecotoxicol. Environ. Saf.* 74, 2099-104.

Joseph, P., 2009. Mechanisms of cadmium carcinogenesis. *Toxicol. Appl. Pharmacol.* 238, 272–9.

Kanter, M., Aksu, B., Akpolat, M., Tarladacalisir, Y. T., Aktas, C., Uysal, H., 2009. Vitamin E protects against oxidative damage caused by cadmium in the blood of rats. *Eur. J. Gen. Pract.* 6, 154-60.

Karmakar, R., Banik, S., Bandyopadhyay, S., Chatterjee, M., 1998. Cadmium-induced alterations of hepatic lipid peroxidation, glutathione S-transferase activity and reduced glutathione level and their possible correlation with chromosomal aberration in mice: a time course study. *Mutat. Res.* 397, 183-90.

Klaassen, C.D., Liu, J., Choudhuri, S., 1999. Metallothionein: an intracellular protein to protect against cadmium toxicity. *Annu. Rev. Pharmacol. Toxicol.* 39, 267–94.

Klaassen, C.D., Liu, J., Diwan, B.A., 2009. Metallothionein protection of cadmium toxicity. *Toxicol. Appl. Pharmacol.* 238, 215–20.

Liu, J., Qu, W., Kadiiska, M.B., 2009. Role of oxidative stress in cadmium toxicity and carcinogenesis. *Toxicol. Appl. Pharmacol.* 238, 209–14.

Liu, T., He, W., Yan, C., Qi, Y., Zhang, Y., 2011. Roles of reactive oxygen species and mitochondria in cadmium-induced injury of liver cells. *Toxicol. Ind. Health* 27, 249-56.

Lowry, O.H., Rosenbrough, N.J., Farr, A.L., Randall, R.J., 1951. Protein measurement with the folin phenol reagent. J. Biol. Chem. 193, 265–75.

Matović, V., Černak, I., Plećaš, B., Stojanović, Z., 1998. Akutno trovanje kadmijumom i oksidativni stres u plazmi kunića. *Jugoslov. Med. Biohem.* 17, 326–27.

Matović, V., Buha, A., Bulat, Z., Đukić-Ćosić, D., 2011. Cadmium toxicity revisited: focus on oxidative stress induction and interactions with zinc and magnesium. *Arh. Hig. Rada Toksikol.* 62, 65-76.

Matović, V., Plamenac Bulat, Z., Đukić-Ćosić, D., Soldatović, D., 2010. Antagonism between cadmium and magnesium: a possible role of magnesium in therapy of cadmium intoxication. *Magnes. Res.* 23, 19-26.

Matović, V., Buha, A., Bulat, Z., Đukić-Ćosić, D., Miljković, M., Ivanišević, J., Kotur-Stevuljević, J., 2012. Route-dependent effects of cadmium/cadmium and magnesium acute treatment on parameters of oxidative stress in rat liver. *Food Chem. Toxicol.* 50, 551-7.

Misra, H.P., Fridovich, I., 1972. The role of superoxide anion in the autooxidation of epinephrine and a simple assay for superoxide dismutase. *J. Biol. Chem.* 247, 3170-5.

Moulis, J., 2010. Cellular mechanisms of cadmium toxicity related to the homeostasis of essential metals. *BioMetals* 23, 877-96.

Nawrot, T. S., Van Hecke, E., Thijs, L., Richart, T., Kuznetsova, T., Jin, Y., Vangronsveld, J., Roels, H.A., Staessen, J. A. 2008. Cadmium-related mortality and long-term secular trends in the cadmium body burden of an environmentally exposed population. *Environ. Health Persp.* 116, 1620-8.

Nemmiche, S., Chabane-Sari, D., Guiraud, P., 2007. Role of α-tocopherol in cadmium-induced oxidative stress in wistar rat's blood, liver and brain. *Chem.-Biol. Interact.* 170, 221-30.

Nordberg, G.F., 2009. Historical perspectives on cadmium toxicology. *Toxicol. Appl. Pharmacol.* 238, 192-200.

Nordberg, M., Nordberg, G., 2000. Toxicological aspects of metallotionein. *Cell Mol. Biol.* 46, 451-63.

Nordberg, G.F., Nogawa, K., Nordberg, M., Friberg, L.T., 2007. Cadmium. In: Nordberg, G.F., Fowler, B.A., Nordberg, M., Friberg, L.T. (Eds.), Handbook on the Toxicology of Metals. Third edition. Academic Press/Elsevier, Amsterdam and Boston, pp. 446–86.

NTP. 2011. Report on carcinogens. Twelfth edition. National Toxicology Program. U.S. Department of Health and Human Services, Public Health. http://ntp-server.niehs.nih.gov/ntp/roc/twelfth/roc12.pdf February 22, 2012.

Plamenac Bulat, Z., Đukić-Ćosić, D., Đokić, M., Bulat, P., Matović, V., 2009. Blood and urine cadmium and bioelements profile in nickel-cadmium battery workers in Serbia. *Toxicol. Ind. Health* 25, 129-35.

Plamenac, Z., Đukić, D., Matović, V., 2002. Contribution to the investigation of the effect of cadmium on lipid peroxidation: an experimental study. *Arch. Toxicol. Kinet. Xenobiot. Metab.* 10, 57–8.

Sarkar, S., Yadav, P., Trivedi, R., Bansal, A.K., Bhatnagar, D., 1995. Cadmium-induced lipid peroxidation and the status of the antioxidant system in rat tissues. *J. Trace Elem. Med.* 9, 144–7.

Silva, N., Peiris-John, R., Wickremasinghe, R., Senanayake, H., Sathiakumar, N., 2012. Cadmium a metalloestrogen: Are we convinced? *J. Appl. Toxicol.* 32, 318-32.

Valko, M., Morris, H., Cronin, M.T.D., 2005. Metals, toxicity and oxidative stress. *Curr. Med. Chem.* 12, 1161-208.

Valko, M., Rhodes, C.J., Moncola, J., Izakovic, M., Mazura, M., 2006. Free radicals, metals and antioxidants in oxidative stress-induced cancer. *Chem.-Biol. Interact.* 160, 1-40.

Waalkes, M.P., 2003. Cadmium carcinogenesis. Mutat. Res. 533, 107–20.

Waisberg, M., Joseph, P., Hale, B., Beyersmann, D., 2003. Molecular and cellular mechanisms of cadmium carcinogenesis. *Toxicology* 192, 95-117.

Witko-Sarsat, V., Friedlander, M., Capeillere-Blandin, C., Nguyen-Khoa, T., Nguyen, A.T., Zingraff, J., Jungers, P., Descamps-Latscha, B.. 1996 Advanced oxidation protein products as a novel marker of oxidative stress in uremia. *Kidney Int*, 49:1304-13.

Yalin, S., Ozlen Sahin, N., Comelekoglu, U., Yalin, E., Bagis, S., Hatungil, R., Dincer Kaya, F. N., 2005. Chronic cadmiun exposure affects the antioxidant defense system in heart of ovariectomized rats. *Pharmacia* 52, 114-7.

Yang, Z., Yang, S., Qian, S.Y., Hong, J.S., Kadiiska, M.B., Tennant, R.W., Waalkes, M.P., Liu, J., 2007. Cadmium-induced toxicity in rat primary mid-brain neuroglia cultures: role of oxidative stress from microglia. *Toxicol. Sci.* 98, 488–94.

Index

A

B

C

D

G

H

I

J

K

L

M

N

O

Q

R

S

T